WHO OWNS THE WATER?

Edited by
Klaus Lanz
Lars Müller
Christian Rentsch
René Schwarzenbach

With the support of EAWAG, the Swiss Federal
Institute of Aquatic Science and Technology

LARS MÜLLER PUBLISHERS

THE WONDER CALLED WATER

16   Water–the Earth's blood

24   Water is everywhere–
the universe is moist

41   Water governs the climate–
Earth's air-conditioning system

55   Water cycles–water wheels,
conveyor belts, and figure
eights

71   Water is different

88   Water and its companions

WATER AND PEOPLE

114   Agriculture, irrigation,
and civilization

129   Water for feeding the hungry

149   Dams and irrigation systems–
temples of modernity

170   Groundwater is running dry

189   Irrigation, soil, and salt

205   In search of solutions

212   Water for people

227   Megacities–the quiet disaster

246   Filth, excreta, and flying toilets

268   How much water
does a person need?

278   Poisoned waters

288   Insidious poisoning

302   Russian roulette

311   Chemicals policy–
sustainable disappointment

320   Hydropower–a mixed blessing

337   The true cost of hydropower

354   Small is beautiful

363 **CLOSE-UPS**

414 **IN THE BEGINNING WAS WATER**

WATER AND POWER

426 Doing business with water

430 Privatization–the
great disappointment

441 Private water supply–
who controls the market?

451 Expensive bottled water

458 On the road to new partnerships

466 Water–a hot political issue

472 Water is power

480 Water and the Middle East
conflict

489 The need for international
water legislation

498 The commercialization
of a human right

510 **WATER BELONGS TO US ALL–
AN APPEAL**

530 Index of countries, rivers,
and lakes

532 Further reading, links

THE WON
CALLED \

DER
VATER

Water–the Earth's blood

Water is wet, tasteless, colorless, doesn't smell, and appears to be rather unexciting. Water seems quite ordinary until you start to wonder about it.

When you do, water becomes mysterious. Why does it rain water and not something else? Why does life not exist without water? Why does water drip in drops and not in threads like other fluids do? Why does ice float on the surface of water when conventional physics dictates that it should sink? Why does water in liquid form seem to occur only on Earth, at least in our solar system? These questions turn water into an interesting puzzle.

Water is different. It has more than forty unusual properties that make it behave differently from other substances. No other substance changes its state so readily and no other substance occurs in nature simultaneously as a solid, liquid, or gaseous material. If water behaved in the same way other substances do, the Earth would not exist today as we know it. Only water becomes lighter when it freezes; all other materials become heavier when they change from the liquid to the solid state. If ice were heavier than water, the Earth would be a desolate planet where life could not thrive. No other substance in nature can store so much energy, transport it for thousands of miles, and release it days or weeks later somewhere else. If this were not so, the Earth would not have a temperate climate but would be covered by icy or burning deserts. Water is a unique solvent and means of transportation for many other chemical substances. If this were not so, there would be no plants, animals, human beings, fertile farmlands, abundant fishing grounds, clean air or a climate in which people could survive.

Water is Earth's blood. It provides nature with everything it needs to survive and it carries away many contaminants that would otherwise disrupt or even destroy nature's balance. Nature would not exist as we know it without water.

H_2O, the most familiar chemical formula in the world, stands for a great mystery, baffling laymen and scientists alike. Indeed, although water is an essential part of our daily lives, we know a lot more about most other substances.

p. 18 Martin Parr/Magnum Photos
p. 20 The Okavango Delta, Botswana. Frans Lanting/National Geographic
p. 22 Fossilized crabs. Hecker/Sauer/Blickwinkel
p. 23 Scan of 20-week-old fetus. gettyimages

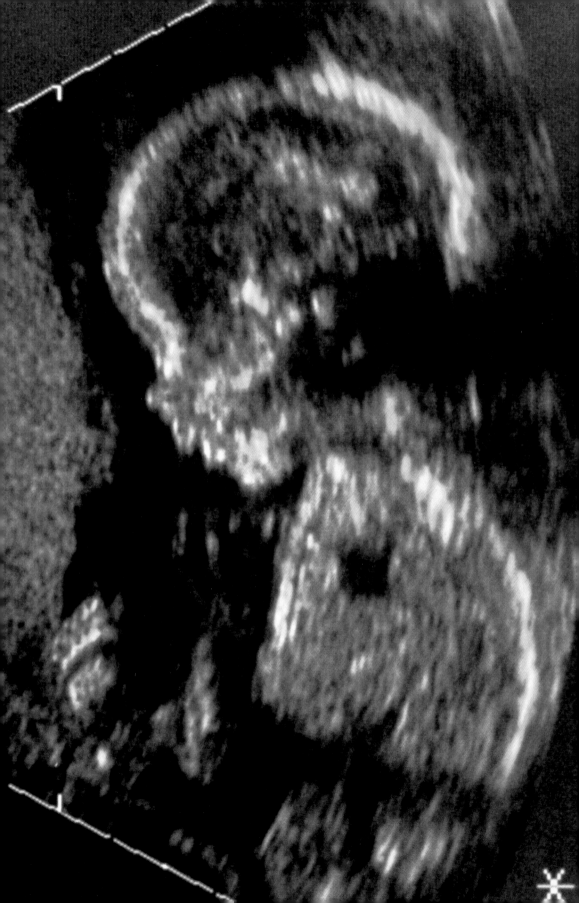

Water is everywhere– the universe is moist

Where does water come from? The long answer to this short question began more than fourteen billion years ago when the universe was born. Only a few seconds or minutes after the Big Bang, when all space, time and matter originated, protons and neutrons, the basic building blocks of matter, took shape from what was an amorphous primordial "soup." In a process that astrophysicists call Big Bang nucleosynthesis, these particles fused, becoming light atomic nuclei. As temperatures cooled from about 1,000,000,000,000,000,000 or 10^{18} degrees centigrade down to a most modest 4,000 degrees, these nuclei, with their electromagnetic fields, captured and held free-floating electrons. The first light chemical elements took shape in this way–hydrogen from one proton and one electron, and helium from two protons, two neutrons and two electrons.

These two elements initially made up 99 percent of the universe's material. Nearly four fifths were hydrogen, one fifth was helium. Shortly afterwards, clouds of this matter massed together as the first stars. Within these nuclear ovens, hydrogen and helium fused into larger structures, creating heavier chemical elements, carbon and oxygen in particular. The whole range of elements assembled in cascades of nuclear reactions and conversion processes and generated the volatile, liquid and solid substances–gases, bubbling magma, metals, rocks– in short, all visible matter. Water, the chemical bonding of hydrogen and oxygen, H_2O, formed in this way long before the seething primordial soup had started to clump together into more recently formed stars.

This process slightly altered the relative abundance of each chemical element. But even today, hydrogen still makes up nearly three fourths and helium one fourth of all matter in the universe. All other elements added together do not make up much more than 1 percent of the total.

The association of two hydrogen atoms and one oxygen atom is favorable in terms of energy and therefore stable, which explains why water has been in existence from the early beginning. Wherever the universe exists, water, H_2O, is there too, whether as single water molecules floating freely in vast, open expanses, or as solid clumps of ice, frozen into rock, as vapor, or as liquid water.

In other words, the universe is full of water. There were huge amounts of water present at some time wherever stars, solar systems and planets have since formed. But this evidently no longer applies today, at least for water as a liquid. Liquid water can occur only on planets with the right temperature regime, neither too hot nor too cold, where average temperatures fall into the small range between the freezing and boiling points of water. Although there may be millions or billions of such planets in the universe, the probability of finding worlds with the right temperatures in any single solar system is very low.

Why? A planet's climate depends basically on two factors–its size and in particular its distance from the sun, the central star of its solar system. If a planet orbits its sun at too great a distance, solar radiation is too weak, the planet cools off and water freezes to ice, creating ice deserts or locking ice crystals solidly into rock. This is evidently what happened to the large majority of planets in our own solar system. But there are exceptions. Photos of the surface of Mars show dried-up riverbeds and valleys, and outlines of seas and oceans. Mars must have had liquid water at some time in spite of its distance from the sun. Astronomers have found a plausible explanation. Apparently the planet's own core, not the sun's energy, once provided enough heat to keep Mars at the right temperature. When its own supply of energy ceased, Mars returned to normal and froze into an icy desert.

When a planet's orbit is too close to the sun, solar radiation heats the planet's surface far beyond the boiling point of water. Temperatures on Venus are around 500 degrees centigrade. Water evaporated long ago and floating gaseous water molecules were blown away into space by solar winds.

Earth, moving in a narrow orbit within our solar system, is obviously the only planet where the sun's energy creates a climate in which water occurs as gas, liquid water, and solid ice at the same time. The juxtaposition of these three states, in particular the presence of liquid water, is indeed the prerequisite for life on Earth.

p. 26 Stars forming in the Eagle Nebula. NASA
p. 27 Halley's Comet. Detlev van Ravenswaay/Keystone
p. 28 Mangala Valles on Mars. ESA/DLR/FU
p. 29 Photo taken by the Galileo space probe, 1992. NASA

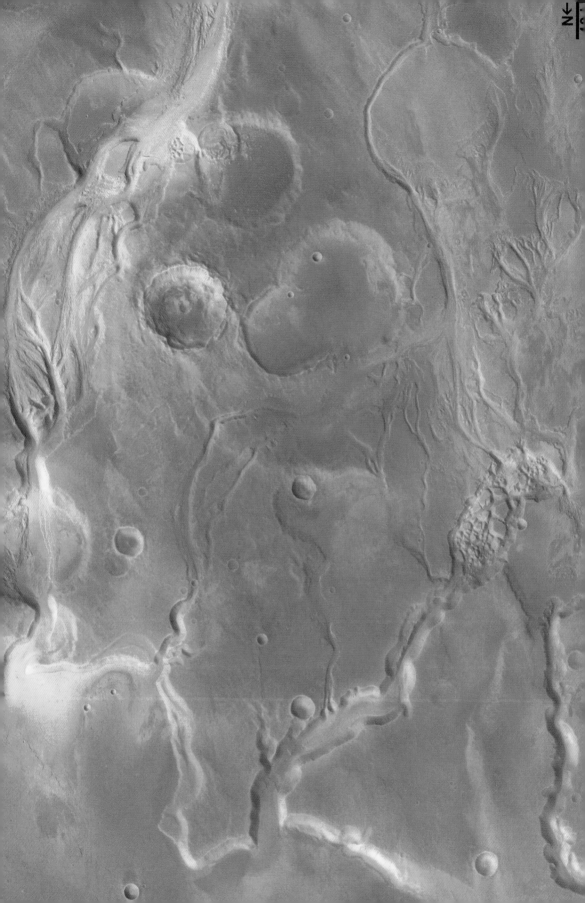

Matthäus Merian the Elder: "Creation", 1625/27. AKG

Sandro Botticelli: "The Birth of Venus", ca. 1482. AKG

Distribution of Earth's Water

Water accounts for three percent of the Earth's mass

1   2   3   4

1	2	3	4
Water	Water at or close to the surface	Freshwater	Liquid freshwater
99% rock under ground 1% at or close to the surface	96.5% oceans 2.5% freshwater 1% saline water on continents	65% ice 35% liquid freshwater	> 99% groundwater < 1% lakes / rivers

Water volumes of the largest lakes

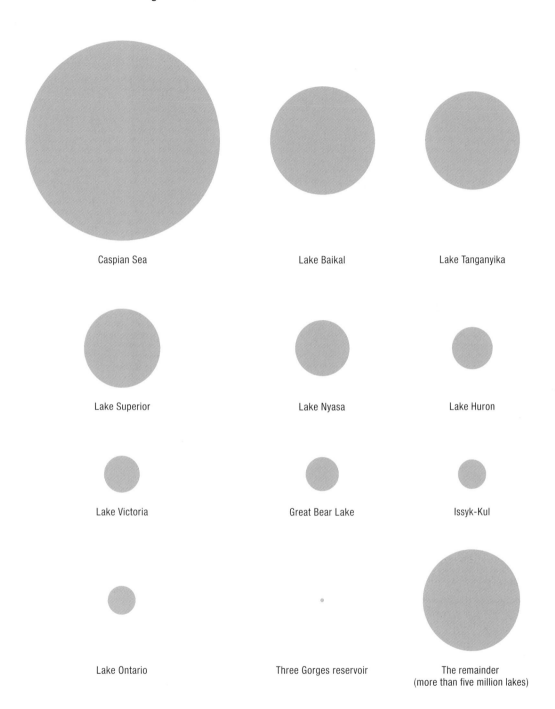

Caspian Sea

Lake Baikal

Lake Tanganyika

Lake Superior

Lake Nyasa

Lake Huron

Lake Victoria

Great Bear Lake

Issyk-Kul

Lake Ontario

Three Gorges reservoir

The remainder
(more than five million lakes)

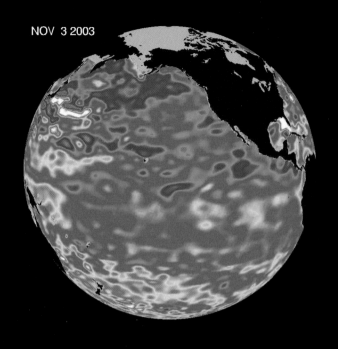

NOV 3 2003

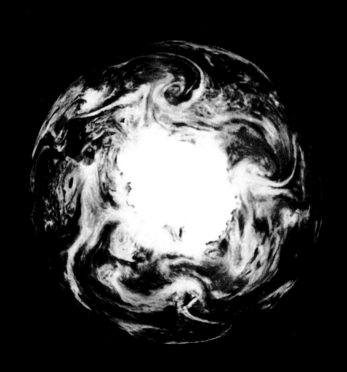

Water governs
the climate

Life as we know it can't exist without water. Our living, organic environment needs water, especially in liquid form, but also as a gas, and as ice and snow. A good thing that water, in contrast to other substances, has an unusual property—not only is it present in great abundance in all three states in Earth's temperate climate, it can also readily change its state without any human involvement. It melts, evaporates, condenses, and freezes. If these changes were not going on all the time, Earth would be a very inhospitable planet indeed, with extreme variations in temperature and an average global temperature of about minus 15 degrees centigrade.

The hydrological cycle is Earth's heating and air-conditioning system. It is what keeps average annual temperatures clearly above the freezing point almost everywhere in the world and sustains global temperatures within a relatively narrow range between minus 40 and plus 40 degrees centigrade. The hydrological cycle moderates differences in temperature not only between warmer and cooler regions, but also between warmer and cooler seasons.

This heating effect, to a great extent self-regulatory, goes on because water can absorb large amounts of solar energy, convert this into heat, store it in the oceans and the atmosphere, and then slowly release it to the surroundings. This complex process takes place during phase transitions when water changes its state by evaporating, freezing, or condensing. The peculiar thing about this process is that when water evaporates, it absorbs large amounts of energy without perceptibly changing its temperature. The energy stored in this way—scientists call it latent heat of evaporation—does not change into heat we can feel until water vapor condenses back into liquid water.

If water did not have this distinctive ability to store heat, making it different from most other substances occurring in nature, about half of the energy radiated to Earth from the sun would be reflected back into space and lost as a source of heat. The remaining energy alone would not be nearly enough to warm the planet above the freezing point. It is water's ability to absorb and latently store energy that means not just half, but about three fourths of solar radiation is captured on Earth. It is precisely this additional one fourth that raises the Earth's average temperature above the freezing point.

Other gases in the atmosphere such as carbon dioxide, methane, and nitrogen oxide also absorb solar energy, but the resultant heating effect is much lower. Water is by far the most important greenhouse gas, responsible for about 60 percent of the natural greenhouse effect on Earth.

This is surprising if we keep in mind that the share of water vapor in the air causing this effect is very small. Depending on temperature and humidity values, this share lies between no more than 0.1 and 4 percent. Equally surprisingly, the total amount of water vapor circulating in the atmosphere is not more than one thousandth of 1 percent of all the water on Earth's surface. These nearly negligible figures are deceptive, however, when we look at the absolute numbers for the gigantic amounts of water involved. Every day, solar energy causes more than 1,300 cubic kilometers of water to evaporate, which adds up to more than 500,000,000,000,000 or $5 \cdot 10^{14}$ tons of seawater every year. Another 74,000,000,000,000 tons or $7.4 \cdot 10^{13}$ of water vapor are released each year from plants and Earth's damp surfaces. At any rate, more water evaporates in less than one week than all of humanity uses during one year.

The mechanism of evaporation has another extremely important consequence for life on Earth—it is responsible for water being clean. If the evaporation process did not take place, there would be no fresh water on the planet. Water everywhere would be as salty as the sea.

If we imagine that water in the oceans and in the atmosphere acts as Earth's heat-storage facility, the hydrological cycle is like a heating system that distributes stored thermal energy all around the globe. Air currents and meteorologically high and low pressure areas with their winds and storms transport stored thermal energy in water vapor to cooler regions. When air cools, it loses its ability to absorb water. Surplus water rains down and releases its stored energy as heat to the surroundings.

The amount of energy that water transports around the globe is enormous, adding up each day to more than seven times the total amount of energy produced by mankind during a whole year. Only one fourth of evaporated water reaches the continents—three fourths rain back down to the oceans.

The water cycle is not just Earth's heating system. It is also a complete air-conditioning system, warming the planet while also, at the same time, preventing it from overheating. Clouds in particular play an important role in this extremely complex, self-regulating mechanism by creating a variable sunshade that protects Earth from too much solar radiation. Clouds reflect a large share of solar energy back into space. This engineers a regulatory cycle that no human could have better invented—when temperatures go up on Earth, more water evaporates and the cloud cover, like a heat shield over the planet, enlarges and effectively moderates further heating of Earth's surface. When temperatures go down, the shield shrinks and allows more direct solar radiation to pass through—and the surface of the planet warms up again.

Many of these feedback mechanisms haven't been fully explained yet. Scientists have not agreed on whether this air-conditioning system really works perfectly and has kept Earth's climate in balance for thousands of years, or whether this balance is just an episode in Earth's history over billions of years. At any rate, it is more or less undisputed that the greenhouse effect caused by human civilization has massively influenced natural heat regulation. Although the greenhouse gases released as a result of human activity have low direct impact on heating, even a minimal increase in air temperature influences the global water cycle, which in turn has an effect on the global climate several times over.

In the same way that liquid water and water vapor function in heating and heat storage, ice and snow are active as cooling and cold-storage systems. When snow and ice thaw each spring, melting consumes large amounts of heat, and this prevents the planet from heating up too quickly during each hemisphere's summer period. At least in temperate zones, this effect moderates differences in temperature over the year. Like clouds, snow and ice reflect back about four times more solar energy than do surfaces on the planet that are free of snow and ice.

Even though our lives are constantly and intricately bound to the water cycle, the cycle remains a rather obscure process, exceeding all powers of imagination and easily tempting us to underestimate the dramatic consequences of even the smallest disruptions and shifts within its complex mechanics. As yet, no climate model can estimate, for example, how the partial melting of water now captured in polar ice caps, glaciers, permafrost, and eternal snow will affect the Earth's water cycle, or global climate or even the water balance in single geographic regions. Numerous amplifying and moderating feedback effects are interconnected in such a complex way that different developments are all plausible. The only certainty is that even relatively low-scale changes in global conditions could massively disrupt and adversely affect the freshwater cycle. The amount of water captured in ice and snow alone is more than 150 times the total amount of fresh water that circulates between oceans and continents. Fresh water altogether makes up only 2.5 percent of the total volume of water on our planet. Of this 2.5 percent, only 0.4 percent keeps the daily freshwater cycle in motion. In short, the fate of the world's population depends on one ten thousandth of Earth's total volume of water. The way in which this minute share of water is distributed over the continents of the planet determines where aridity, fertility, desolation and disastrous floods will occur. Whether the world's population has enough clean drinking water and farmers have enough water to irrigate their fields will depend on how much, and equally importantly, how clean, this minimal amount of water is.

p. 36 Dusky Sound, New Zealand. Gordon W. Gahan/National Geographic
p. 37 Windeck/vario-press
p. 38 Larry Gatz/gettyimages
p. 39 Greina plateau, Switzerland. Arno Balzarini/Keystone
p. 40 Surface temperatures in the Pacific: Red indicates warm and blue cool temperatures,
on a long-year average. NASA/JPL Ocean Surface Topography Team
p. 40 View of the South Pole from outer space: The area that appears dark at night was eliminated
by combining several photos taken by the Galileo space probe. NASA
p. 45 Laurent Gillieron/Keystone
p. 45 Martin Parr/Magnum Photos
p. 46 Georgios Kefalas/Keystone
p. 47 Széchenyi Baths, Budapest. Martin Parr/Magnum Photos
p. 47 Robert Ghement/Keystone/EPA Photo

Precipitation: global distribution

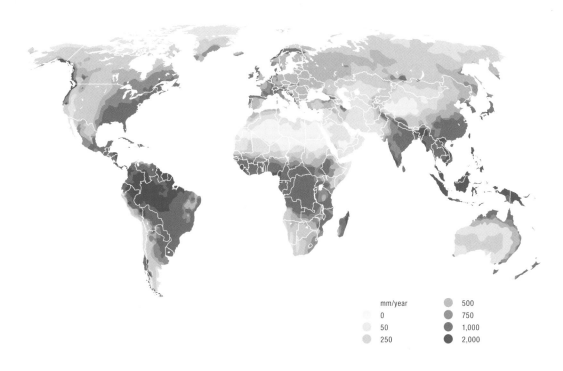

mm/year			
0		500	
50		750	
250		1,000	
		2,000	

Groundwater: global distribution

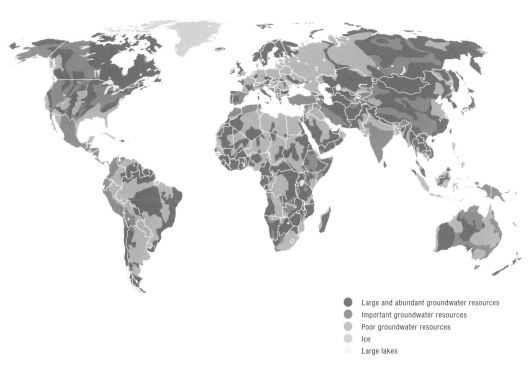

● Large and abundant groundwater resources
● Important groundwater resources
● Poor groundwater resources
● Ice
● Large lakes

Water scarcity: prediction for 2025

- No or little water scarcity
- Economic water scarcity
- Physical water scarcity
- No data available

What happens to the precipitation?

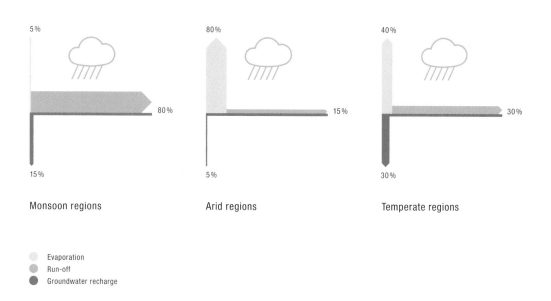

| Monsoon regions | Arid regions | Temperate regions |

5% · 80% · 15%

80% · 15% · 5%

40% · 30% · 30%

- Evaporation
- Run-off
- Groundwater recharge

Water cycles–water wheels, conveyor belts, and figure eights

The data processing ability of all the world's computers together is not powerful enough to simulate a roughly accurate picture of the water cycles occurring in even a small place like a village. If it were feasible, a computer simulation would look like an endlessly huge machinery in which individual parts, in enormously complicated ways, are interconnected, mesh with one another, exert mutual influence, drive each other forward and apply brakes. Water cycles link processes taking hours, days and weeks to other processes taking years, decades, or even millions of years. They determine each day's weather, balance the exchange of heat between different regions of the world when seasons change, and regulate the change between ice ages and warmer periods over thousands of years. Over much longer periods of time, they even shift Earth's tectonics. Water cycles transport immense amounts of water over thousands of miles at the same time they exhibit microscopic activity in plant, animal and human cells. They have been instrumental in shaping the Earth's surface, creating mountains and valleys, rivers and seas, and within a period of three-and-a-half billion years they have enabled the development of life on Earth. Water is, at the same time, the medium in which new life starts billions of times every second.

In an everlasting journey lasting millions and billions of years, every molecule of water has moved in countless cycles. As part of a raindrop that just fell from the sky to our doorsteps, a molecule might have been swimming in the ocean just a few days earlier. Maybe the Gulf Stream carried it across the Atlantic Ocean from the Caribbean after it spent several hundred years circling between Africa and the coast of Latin America. Maybe it spent decades in smaller cycles in Brazilian rainforests. Before that, it might have been in the Indian Ocean, joining monsoon rains for years or thousands of years or it might have drifted in the North Pacific Stream between California and Japan after spending hundreds of thousands of years captured in the ice of Greenland. Originally, a volcano might have hurled it to the surface after it had been trapped in the Earth's crust for several billion years.

A molecule of water can go on its never-ending journey only because water effortlessly changes its state and therefore can move easily from one cycle to another. The hydrological cycle is the most fundamental of them all (→ p. 64). Many other cycles wouldn't function without it. It links cycles in the ocean with those on land. It is the life-support base of the whole biosphere. It plays this fundamental role, however, only because it is closely coupled with two oceanic cycles that make the water system a global one, spanning the whole planet.

These ocean cycles or current systems, one on the surface and the other in the oceans' depths, are the real conveyor belts that regulate the exchange of heat around the world, whereas the hydrological cycle is the agent behind the fine distribution of heat over single continents.

Circulation on the surface gets the world's large oceans to move–the Atlantic, Pacific, and Indian Ocean (→ p. 65). At most not quite 100 meters deep, these currents are driven mostly by winds and large global wind systems. The Earth's rotation also plays an important role. The effect of rotation on the circulation of water is called the Coriolis Effect after the French mathematician and physicist Gaspard Gustave de Coriolis. Planetary rotation induces eddies which rotate clockwise in the southern hemisphere and counter-clockwise in the northern hemisphere. The Coriolis Effect generates two ocean-wide eddies, one in the northern, the other in the southern hemisphere, in each of the world's three large oceans–the Atlantic, Indian, and Pacific Ocean. These six eddies include the Gulf Stream in the North Atlantic, which influences the European climate, the Agulhas Current off the coast of southeast Africa, and the Kuroshio Stream off Japan, which as it moves across the Pacific also influences the climate of the Californian coast.

The intensity of these oceanic currents varies according to the season, causing currents to exert great influence on the climate and weather of the continental regions they flow past. The North Equatorial Stream in the Indian Ocean is the driving force behind monsoon rains, even changing its direction during the year. It flows clockwise from June until September, counter-clockwise from November until March, and generates drastic changes in weather conditions in India, Bangladesh, and Pakistan.

These six regional eddies are linked around the world by the deep circulation (→ p. 65) that exchanges water in figure-eight-like conveyor belts flowing around the Earth through the Atlantic, Pacific, and Indian Ocean. These conveyor belts keep water at all depths of the oceans in constant motion. Called thermohaline circulation by scientists, this drift pushes warmer, lighter water in the Atlantic in a northerly direction. Evaporation causes the current to become saltier, and as heat is released, it becomes cooler until it gradually sinks down around Greenland. At a depth of as much as 1,000 meters, the current flows back south. The warmer northerly current above prevents it from rising so that, as a cold deep current in the South Atlantic, it has no other way to flow than into the Indian Ocean from where it continues into the Pacific. Here it rises to the surface and warms up in a loop in the South Pacific, and then south of Africa flows back into the Atlantic.

This entwined and complicated mechanism of water cycles is possible not just because water changes its state so readily. Water has other very unusual properties too. No other comparable substance occurring in nature that might be able to perform similar functions has a boiling point as high as water's. Given Earth's prevailing temperatures, any of these other substances would evaporate much more quickly than water does. Not only that, no other substance is lighter as a solid than as a liquid. These two properties enable the existence of water cycles. If ice were heavier than water, it would sink to the bottom of lakes and oceans. Lakes would freeze at the bottom and exterminate fish and plants. Lakes and maybe even the oceans would have frozen from the bottom up over millions of years. What seems to be the most ordinary and normal process in the world happens only because water doesn't behave in a "normal" way—because water is different from everything else.

Diagram of Meteorology, 1846. Science & Society Picture Library

Willy Stöwer: "Sinking of the Titanic", 1912. AKG

Heat budget of the Earth

Annual average in watts per square meter

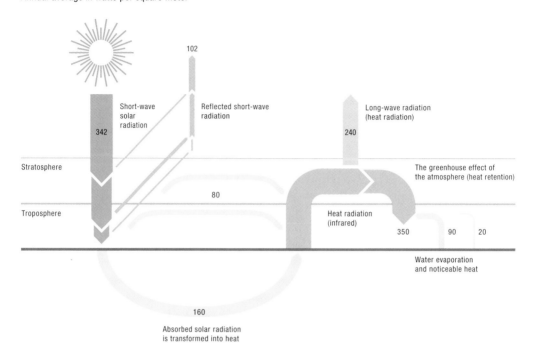

102

342   Short-wave solar radiation

Reflected short-wave radiation

Long-wave radiation (heat radiation)

240

Stratosphere

The greenhouse effect of the atmosphere (heat retention)

80

Troposphere

Heat radiation (infrared)

350   90   20

Water evaporation and noticeable heat

160

Absorbed solar radiation is transformed into heat

The water cycle

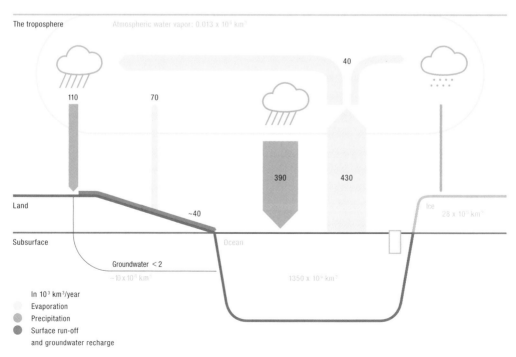

The troposphere   Atmospheric water vapor: 0.013 x 10⁶ km³

40

110

70

390

430

Land

Ice
28 x 10⁶ km³

Subsurface   Ocean

~40

Groundwater < 2
~10 x 10⁶ km³

1350 x 10⁶ km³

In 10³ km³/year
Evaporation
Precipitation
Surface run-off
and groundwater recharge

Water residence times in aquifers

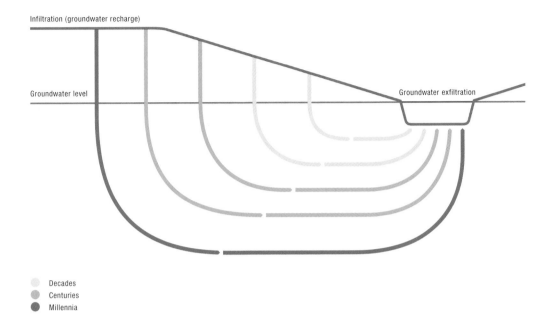

Infiltration (groundwater recharge)

Groundwater level

Groundwater exfiltration

- Decades
- Centuries
- Millennia

Global ocean circulation

Ocean surface currents (warm, low salinity, light)
Deep water currents in the ocean (cold, high salinity, high density)

p. 66 Esmeraldas, Ecuador. Alex Webb/Magnum Photos
p. 67 Atacama Desert, Chile. Pierre Hausherr/laif
p. 68 Lake Aral, Kazakhstan. Gerd Ludwig/Visum
p. 69 New Orleans. Justin Sullivan/gettyimages
p. 75 Central Park, New York City. Thomas Hoepker/Magnum Photos
p. 76 Gianni Cigolini/gettyimages
p. 84 Sahara, Algeria. Gunther Michel/BIOS
p. 85 Northwest Territories, Canada. Bryan Cherry Alexander/laif
p. 86 Ferdinando Scianna/Magnum Photos
p. 86 Mammoth Cave National Park. Kentucky Adam Jones/Photo Researchers
p. 87 Fes, Marocco. Abbas/Magnum Photos
p. 87 Grand Canyon, Arizona. Hiroji Kubota/Magnum Photos

Water is different

Water fools us. The most common and apparently most ordinary and normal liquid on Earth is actually the most unusual one. In many ways it behaves differently from what we would expect, certainly differently from other "normal" liquids.

It's not easy to explain why. Water's secrets are buried in its molecular structure, an area not open to scrutiny by human senses. Almost all the ideas we have about water are based on theoretical models and simplified pictures that can't fully show what's really important in enabling us to understand it.

The most common diagram of a water molecule in the sign language of chemistry, as we all remember from our school books–an isosceles triangle with one oxygen atom and two hydrogen atoms–hardly helps us identify water's unusual properties.

Two other simple pictures of the water molecule aren't of much help either. One three-dimensional model shows the oxygen atom as a large ball with two smaller hemispheres of hydrogen glued to its surface. The second model shows the water molecule as a tetrahedron with the oxygen nucleus in the center and the four corners made up of two hydrogen atoms and two free electron pairs.

But all these pictures can't show us that a molecule of water is not a rigid object but rather a dynamic and very agile entity. Up until a few decades ago, scientists therefore used a different model to help their understanding, imagining that a molecule was like a tiny planetary system in which electrons orbited around atomic nuclei, with its individual celestial bodies attracting and repelling each other.

Although quantum physics has proven this model to be wrong, at least this false yet graphic model allows us to demonstrate properties that also apply to the correct but no longer comprehensible model defined by quantum physics. We can imagine a molecule of water as a kind of planetary system that is in an unstable and rather restless balance. Then we can imagine water as an association of numerous planetary systems that are also in a state of mutual attraction and repulsion, and that share some planets. If energy is added to this construction, individual parts start to vibrate. As more energy is added, parts vibrate and rotate more strongly and constantly. In this way, a molecule of water absorbs radiation and converts it into kinetic energy. If molecules get into cooler surroundings, movements slow down and become weaker. Surplus energy is released and changes into heat, which is released into the surroundings.

When the temperature is close to the freezing point, these movements don't stop altogether, but single parts move so minimally that the whole construction hardens and becomes rather stable. The loose association of water molecules solidifies into snow or ice crystals.

In this situation however, molecules of water don't react in the same way towards all kinds of energy. Energy from microwave and infrared radiation will generate the most molecular movement—these types of radiation are absorbed. Other kinds of energy, such as rays of visible light, leave water molecules unaffected. They slip through water molecules without having much effect. If water molecules absorbed visible rays the way minerals and metals do, Earth would be a very dark place indeed.

But even this concept of water molecules leaves many questions unanswered. Other comparable molecules react similarly to water molecules when energy is added—they start to move, and, in turn, release this kinetic energy when it gets cooler. What is so different and unusual about water isn't evident until we look more closely at the structure of a water molecule.

Every atom consists of a nucleus of protons and neutrons wrapped in a cloud of electrons. As a rule, an atom has exactly the same number of negatively charged electrons as positively charged protons, while neutrons have no electrical charge. Since the structure of atoms can't really be depicted graphically but expressed only in mathematical probabilities, we will return here for practical purposes to the planetary system model, which is wrong in principle, but at least something we can visualize.

In this model, we perceive the electrons whirling around the atom's nucleus as relatively inconstant companions. Some of them are easily lured away and captured by other nuclei. Atoms often balance out such a loss by getting together with other atoms, sharing and using electrons mutually. In this case, we should imagine an electron not as a single whirling particle but as a kind of cloud wrapped around the nucleus in a diffuse structure.

Electrons have a special liking for joining together in pairs. Hydrogen is an ideal partner that likes to do this. Since it is the lightest of all chemical elements, hydrogen is attracted by heavier atoms more easily than other elements. Hydrogen has only one electron, which predestines it for enticement as a "single." These properties make hydrogen a very sociable element.

The oxygen atom's outermost electron shell (next to two electron clouds filled with pairs) has two electron clouds which have only one electron each. Quantum physics dictates that the electron clouds of light elements must have double occupancy. This physical law makes oxygen react chemically in the effort to fill electron-deficient spaces. Accordingly, an oxygen atom gladly joins up with two hydrogen atoms, each possessing a single electron. Since the four electron clouds resulting from this union mutually repel each other because of their negative electrical charges, they take up positions as far from each other as possible on the round shell encircling the oxygen nucleus. They form the tetrahedron typical for a molecule of water. Two free electron pairs are on two corners of the tetrahedron, while each hydrogen atom is docked onto the other two corners where the single electrons of hydrogen and oxygen have joined. This tetrahedral arrangement affects the distribution of electrical charges within the whole molecule. Since both hydrogen electrons are oriented towards the oxygen atom, a slight positive charge forms on their exterior, while the two free electron pairs exhibit a slight negative charge. This highly uneven distribution of charge gives water molecules their strongly dipolar property in comparison to similar molecules.

This dipolarity shows its effect especially where many water molecules group together. They build what scientists call hydrogen bridges between single molecules, and thus attract each other strongly (→ p. 82). This is the real secret behind many of water's unusual properties. Practically no other molecules in existence possess these characteristics in similar form and expression.

The strong mutual attraction across hydrogen bridges explains why water has such high freezing and boiling points. The molecules of other light substances are much less attracted to each other and therefore break apart at temperatures much lower than 100 degrees centigrade. It requires much more energy, however, to overcome the attraction between water molecules. When ice melts, a lot of energy must be expended to make hydrogen bridges movable to start with, and finally, to get them to break apart when water vaporizes. Even then, single bridges remain in existence, which is why water vapor still forms the tiniest of droplets. Hydrogen bridges don't all break completely apart until temperatures reach about 1,000 degrees centigrade.

Of course, this graphic description is only an incomplete picture of reality. The seemingly solid hydrogen bridges we have imagined are in fact only statistical probabilities. In other words, the real life span of a hydrogen bridge is no more than one billionth of a fraction of a second. Bridges permanently break apart and form again, but in their entirety they are more or less stable on statistical average. The short-livedness of single (real) hydrogen bridges means that the four poles of a water molecule are not all anchored to a bridge at the same time. On average, it is only two or three that are.

The phenomenon of hydrogen bridges also explains two other unusual properties of water. To a certain degree, each individual molecule of water behaves no differently from the molecules of all other substances. When temperatures drop, molecular movement becomes more sluggish and occupies less space. Single molecules move closer together and their grouping becomes denser and heavier. The transitory nature of hydrogen bridges shakes molecules of liquid water again and again in different ways, bringing them closer together. But this rule, applying to other substances as well, is valid for water only at temperatures above 4 degrees centigrade. In a situation where other molecules stack together more and more densely during the transition from the liquid to the solid state, hydrogen bridges cause water molecules to keep their distance from each other. When ice is formed, they set into a relatively wide-meshed lattice. Therefore, in contrast to other substances, water molecules are less densely packed within ice lattices than they are in the liquid state. Oddly, therefore, ice is lighter than water.

Secondly, the dipole character of a molecule of water is also behind another important peculiarity, which is that water drips in drops and not in threads as many others substances in the liquid state do. Water has a very high surface tension. The attractive forces influencing each water molecule are more or less equally strong on all sides in the center of a larger or smaller grouping of water molecules. But things are different on the surface. Here, molecules on the very outside are pulled only from within, and quite strongly so. Forces of attraction work like an electrical fishing net or a skin that holds water together. The property of water (not only water, but watery liquids in general), that its molecules form into drops, is something very special indeed.

All these properties allow water to circulate in all these large and small cycles, which in turn form the basis of any life on this Earth.

Kenneth Wyatt: "Jesus Walking on Water". 2006. Jerusalem Center for Biblical Studies

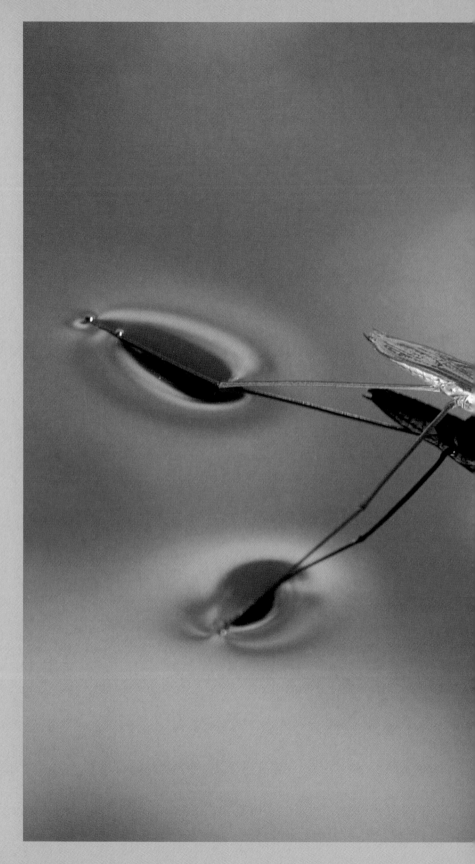

Pond skater. H. Schmidbauer/Blickwinkel

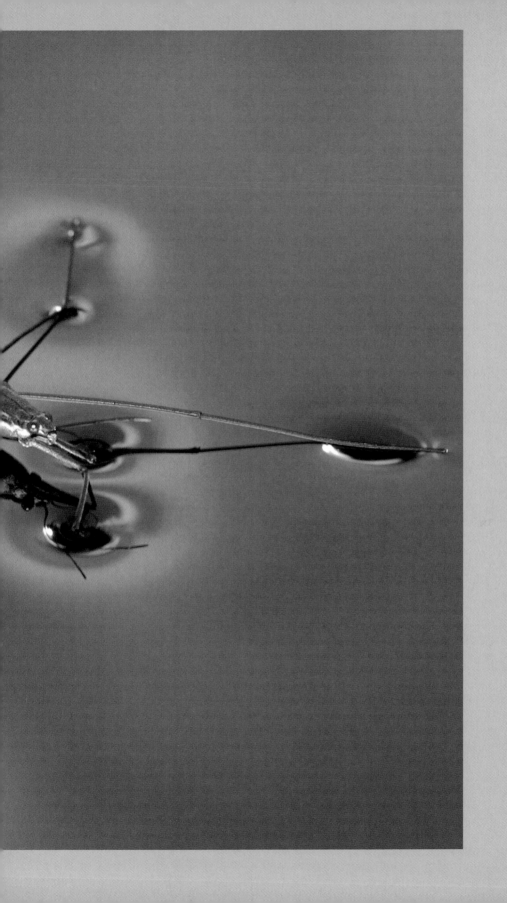

Structure of the water molecule

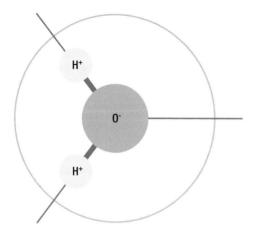

Aggregate states

Gaseous (no interaction)

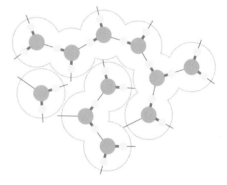

Liquid (dynamic interaction)

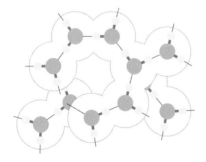

Solid (static interaction)

Hydrogen
Oxygen

Thermal energy content of water

Water can store a lot of thermal energy—however, the energy is stored less in tangible heat (the "temperature") than in latent form in the transitions between the various aggregate states.

Heat supplied per kg water (10^5 J/kg)

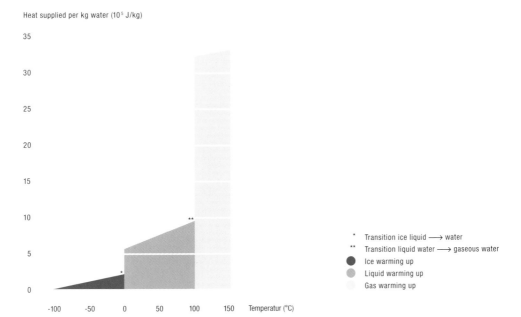

*   Transition ice liquid $\longrightarrow$ water
**   Transition liquid water $\longrightarrow$ gaseous water
●   Ice warming up
●   Liquid warming up
○   Gas warming up

Aggregate states at different temperatures

Other small molecules like H_2O exist in the form of gas under temperatures typical on Earth. Astoundingly, only water appears in this temperature range as ice, liquid and gas.

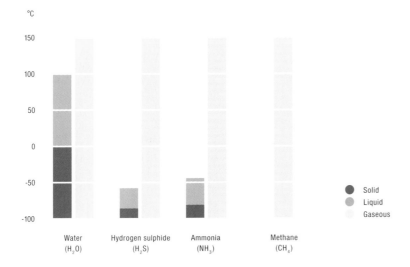

● Solid
● Liquid
○ Gaseous

83

Water and its companions

Water almost never appears by itself. At least in nature, it's almost always accompanied by other substances. These companions give it its character, causing it to smell fresh or like chlorine, taste like Coca-Cola or something rotten, feel hard or soft, soapy or corrosive. Pure water exists almost only in theory or in science labs.

Water transports everything from enormous boulders and infinitesimal molecules to dead material and living bacteria, vital nutrients and highly toxic substances. Water is indiscriminate, carrying on the one hand whatever plants, animals, and humans need to come into existence, grow, and live, and, on the other hand, many things that threaten life, make it ill, poison, and even destroy it.

Its unique role as a means of transportation and as a medium is due to its special property as a solvent for certain chemical substances, especially for (positively or negatively charged) ions that can enter into strong interaction with dipolar water molecules. This includes many elements that are essential for plants, animals, and humans, such as sodium, potassium, magnesium, calcium, and silicon, but also trace metals like iron, copper, and zinc ions.

Water molecules settle their positively or negatively charged sides down on these ions to the point where ions are so completely shrouded that they disappear, so to speak, in water—they dissolve.

But even neutral molecules that can build hydrogen bridges as water molecules do (→ pp. 71), although not in possession of positive or negative charges, can dissolve exceptionally well in water. These include alcohols and sugars. Water does, however, have a hard time dissolving larger non-polar molecules that exert little attractive force on water molecules.

Basically, any substance can be dissolved in water, at least in very small quantities. Ultimately, differences lie only in the degree of solubility. While up to 300 grams of common salt can be dissolved in 1 liter of water, the corresponding figure for the insecticide DDT is only 0.000003 grams –in other words, 100 million times less. But we shouldn't be deceived by this low number. Theoretically, $5 \cdot 10^9$ tons of DDT could be dissolved in the world's oceans. In reality, the figure is of course much lower.

The amount of any substance that can be dissolved in water depends on how strongly the substance's molecules adhere to each other or how rigidly they are bound within a crystal lattice. The more energy expended to separate them, create space between water molecules, integrate the substance's molecules into free spaces and re-group water molecules around them, the less soluble the substance is.

If water isn't able to dissolve any more ions or molecules of a substance, it is saturated. If any more of the substance is added, it won't be dissolved but will drop to the bottom or float as drops, a puddle, or a film on the water's surface. Or settle on another surface, like calcium carbonate does in water pipes. Chemical reactions with other substances can also cause the original substance to precipitate. For example, reduced iron ions dissolved in water will oxidize when they react with oxygen, producing sparingly soluble rust.

Gaseous substances can also exceed the saturation point–they then bubble up to the surface like the CO_2 in mineral water. In August 1986, an enormous gas cloud of toxic carbon dioxide escaped from Lake Nyos in Cameroon, killing 1,700 people and most animals within a radius of 25 kilometers. Scientists now presume that a volcano under the lake had been enriching the water with carbon dioxide at a depth of 200 meters for decades. Water pressure, the weight of 200 meters of water, had originally prevented the gas dissolved in the water from bubbling up to the surface. Scientists assume that the volcano's crater on the lake floor partially collapsed, disrupting the water so much that about one million cubic meters of carbon dioxide were set free within seconds and escaped to the surface. CO_2 has no smell and is heavier than air. It must have suffocated humans and animals in their sleep.

Since then, scientists have discovered several other "killer" lakes that are similarly threatening, such as Rwanda's Lake Kivu, which is 2,000 times larger than Lake Nyos. Some two million people live on its shores. Another danger lurks in these lakes besides the presence of carbon dioxide. Water at the bottom also contains very high concentrations of methane, an explosive gas that is the cause of many mining disasters.

Scientists are now trying to release the dissolved carbon dioxide through pipes in a controlled way. They hope to use the same method to tap the methane gas as well, and use it for energy production. Indeed, methane gas reserves in Lake Kivu could cover all of Rwanda's energy needs for the next 400 years.

In practice, especially in environmental chemistry, it is, however, much less important to know how much of a certain substance theoretically can be dissolved in water. The actual quantity of a substance dissolved in certain bodies of water is of much more interest. In most problem cases, the real concentration is far less than the theoretically possible saturation point (→ p. 96). For example, one milligram of geosmin, a musty-smelling substance produced by microorganisms, is enough to render 100,000 liters of potable water undrinkable. But geosmine's saturation concentration is about 10 million times higher.

In all bodies of water, even in fresh spring water far from any civilization, but particularly in effluent from agriculture and industry, and sewage water from cities and villages, we will find at any time dozens or thousands of different materials and substances, all of them behaving in varying ways. This poses environmental chemists a number of extremely complicated problems. Not all of water's companions are harmless.

Two groups create particular problems: hydrophilic and hydrophobic persistent substances. Hydrophilic (water-loving) substances dissolve easily in water and are therefore very difficult to remove from it, at least in a natural way. One of these is MTBE, an additive to gasoline, of which more than 10 million tons (and this amount is increasing) are consumed each year around the world. We still don't know very much about its long-term effects. The concentration of MTBE in water in some places in California is so high that, although the drinking water supply is not poisonous, water nevertheless smells so foul that it can't be used for drinking.

Conversely, hydrophobic (water-hating) substances use every opportunity to escape from water as soon as they can. They accumulate, for instance, in the fat of living organisms and are then passed through the food chain from plants to animals and humans. Among the most problematic of these substances are the persistent organic pollutants (POPs) such as the insecticide DDT, dioxins, and poly-chlorinated biphenyls (PCBs).

Serious environmental problems also arise, however, when the natural composition of water—for instance, the levels of oxygen, nitrogen, and phosphorus—are so disrupted by external influences that nature is no longer able to deal with them. Enormous amounts of nitrates and phosphates, from excreta as well as from detergents and fertilizers, have led to the massive over-fertilization of many bodies of water. Consequences have been devastating. The more nutrients there are in water, the more rapidly plant and animal plankton reproduce, as do the bacteria that feed off dead organic material. These bacteria consume enormous amounts of oxygen. A biologically fatal program—when oxygen is gradually depleted, the life-support base of other, higher plants and animals in water is destroyed.

Called eutrophication, this phenomenon has heavily damaged numerous lakes since the 1950s, as well as many coastal waters. In wealthier industrial nations, this problem has been reduced to a certain degree thanks to a number of very expensive measures. For instance, the use of phosphates in detergents has been banned. Many efficient sewage-treatment plants have greatly reduced the entry of nitrogen species and phosphates into water. But in spite of international agreements, even today one million tons of nitrogen species and 100,000 tons of phosphates flow into the North Sea alone every year. Even in wealthy industrial nations, we are still a far cry from really solving this problem in a sustainable way.

Water's magnificent ability to dissolve almost all substances, and make them apparently disappear as long as they are diluted enough, has led people to believe for thousands of years in the wonderful illusion that

water has the inexhaustible capacity to clean itself. For thousands
of generations in the past, and in many places even today, people solved
their waste problems in the easiest way imaginable–they dumped most
of their solid and liquid wastes into bodies of water, and buried or burned
the rest so that it all simply disappeared.

Only gradually have we realized that these times are finally over. Industrialization and the population explosion have left us with mountains of
waste and quantities of effluent that are beyond the cleaning capacities
of water and nature. Water and nature are less and less able to clean
themselves. We humans must learn, whether we want to or not, to do this
job ourselves, making sure we don't destroy our own life-support base.

冨嶽三十六景 神奈川沖浪裏

北斎改為一筆

Katsushika Hokusai: Kanagawaoki Namiura, 36 views of Mount Fuji. Keystone/AP/Arthur M. Sackler Gallery

The Sirens, Bienal de Valencia 2005

Substances and their solubility and effect concentration

Substance	Water solubility[a] (mg/l)	Effect concentration	Type of effect
DDT (insecticide)	0.003	6.25 mg/kg[c]	Possibly carcinogenic
Atrazine (herbicide)	30	0.5 mg/kg[c]	Possibly carcinogenic
MTBE (gasoline additive)	50,000	0.02–0.04 mg/l	Bad taste
$CaCO_3$ (chalk, water hardness)	15.3		Calcium deposits (important for detergent dosage)
NaCl (common salt)	36,000	200–300 mg/l	Tastes salty
Benzo[a]pyrene (incineration product)	0.0038		Carcinogenic
Iron (III)(hydr)oxide (rust)	0.0000000001		Brownish precipitation (rost particles)
Iron(II)chloride	64,400	0.04–3 mg/l	Metallic taste
Sodium nitrate	92,100	500 mg/kg[c]	Methaemoglobinaemia, possibly carcinogenic
Geosmine (produced by algae)	150	0.000005–0.00001 mg/l	Earthy/musty taste

Substance	Typical concentration River/lake (mg/l)	Groundwater (mg/l)	Drinking water (mg/l)	Guideline for drinking water[b] (mg/l)
DDT	0.00001–0.00084		< 0.00001	0.001[d]
Atrazine		0.00001–0.006	0.00001–0.005	0.002 (WHO), 0.0001 (EU)
MTBE	< 0.012	< 0.6		USEPA recommendation < 0.02–0.04, California's recommendation 0.013
$CaCO_3$	100–200	≤ 500	10–500	No value
NaCl	2–50	< 500 (seawater infiltration)		< 250 (WHO) taste
Benzo[a]pyrene		0.0003–0.001		0.00001[e]
Iron(III)(hydr)oxide				No value
Iron(II)chloride				0.2 (as iron)(EU) aesthetics
Sodium nitrate	< 20 (as nitrate)	< 1500 (as nitrate)		50 (as nitrate)(EU)
Geosmine	< 0.0002		< 0.00005	No value

a Quantity soluble per litre
b Guidelines for drinking water: WHO, EU, USEPA
c mg/kg body weight (rat)
d Intake by other foodstuffs is more important than by drinking water
e Intake by other foodstuffs is much more important than by drinking water (< 1 %)

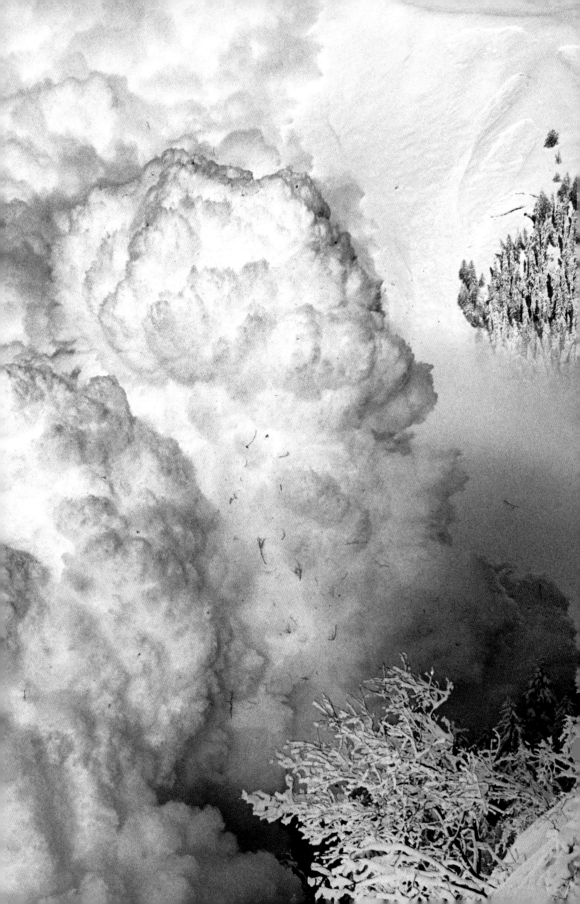

WATER AND PEOPLE

Agriculture, irrigation, and civilization

We can't talk about water without talking about agri-culture. It accounts for 70 percent of all the freshwater consumed by humans. In the Middle East, Africa, and Asia, it even accounts for 80 to 90 percent of all water used.

But we can't talk about agriculture without talking about irrigation. Two out of every 3 tons of grain grow on irrigated fields. Likewise, in many places the cultivation of rice, cotton, and numerous other agricultural products depends on irrigation.

In an effort to satisfy their agricultural industries' growing demand for water, the most populated countries in the world, but also industrialized countries like the United States and Spain, construct ever larger dams, channel rivers over distances of hundreds of kilometers, build extensive irrigation canals, and exploit fossil groundwater reserves.

The industrialization of agriculture, called the green revolution, has indeed greatly improved the food supply for millions of people. But at the same time, industrial methods of irrigation and cultivation have caused serious and sometimes irreversible ecological damage. Cropland is being increasingly contaminated and damaged by fertilizers and pesticides, soil erosion, salinization, and pollution created by human and animal excreta. The stepped-up pace of agricultural industrialization is also partly responsible for a number of incalculable social and economic problems.

For these reasons, the green revolution of the past fifty years should now be followed by a reformed system of agriculture that has food security and the wise use of water resources at its core. Experts refer to this idea as the blue revolution. Such a reformed system would treat nature more carefully and would manage renewable water resources more efficiently. Taking into account social circumstances and economic conditions, especially in developing countries, it would guarantee the survival of the approximately 820 million people still going hungry.

Splendid, but wasteful! Most of the finely sprayed water evaporates before it reaches the ground.
Fruit plantations in South Tyrol. Klaus D. Francke/Bilderberg

When there is insufficient rain for cultivating the land and preserving the countryside, the accessibility of water becomes a question of survival. Mali, Sahel. Abbas/Magnum Photos

Thanks to irrigation (as in Najd, for example) the desert-state of Saudi Arabia has become an important wheat exporter. The groundwater used there, which is drawn from old (fossil) reserves, is barely replenished and will be depleted in a few years' time. Ray Ellis/Keystone

Globally, art
uses about t
much wate
households

ficial irrigation
en times as
as all private

The scarcer the reserves, the more important it is that water be justly distributed: a traditional irrigation water dispenser in the Timimoun Oasis, Algeria. Frans Lemmens/Das Fotoarchiv

Thousands of years ago, the ancient Egyptians worshipped the Nile like a god. Everything depended on its water; without it there would be no green areas in Egypt today. A date plantation in Rashid near Alexandria in the Nile delta.

Stuart Franklin/Magnum Photos

Industrious monuments from an old civilization: The Norias, ancient water-wheels in Hama, Syria, still lift water from the Oronte into viaducts as they did 2,000 years ago. The water is then channeled onto the fertile fields in the surrounding area. Sylvain Grandadam/Keystone

Irrigation has turned Vinschgau, the region with the lowest rainfall in the Eastern Alps, into Tyrol's corn belt. Evidence reveals that from the 12th century on, so-called Waale appeared: irrigation channels that directed water from glaciers and streams, often in a most extraordinary manner, into the fields from remote valleys. Galli/laif

As forests are cut down in Haiti, ever more sources of drinking water are drying up. The Swiss development organisation Helvetas is working together with the population to plant hundreds of thousands of tree seedlings: combining forces at the tree school in Chateau, Grande GrosMorne. Fritz Brugger/Helvetas/Keystone

A woman's job the world over: Mothers and daughters operating a handpump in Chainpur in Nepal. The pump not only supplies the family with water but also waters the vegetable garden. Caroline Penn/panos

Check dams made from rocks are built to store run-off during the rainy season for the dry months. Eghade, Tahoua province, Niger, 2005. Daniel Auf der Mauer/Keystone

Water for feeding the hungry

Rain is an unreliable ally. Sometimes it floods entire regions, ruining crops and destroying towns and villages. Elsewhere it fails to rain just when water is most badly needed. This is why, for more than 9,000 years, people have been storing water during rainy seasons to water their fields during dry periods. They have also learned to transfer water from areas with greater rainfall to more arid regions. The art of irrigation accompanied the rise of civilization. Together with language, it was one of humanity's first great cultural accomplishments. Irrigation requires a high degree of technological know-how and skill, and more importantly, a degree of collective organizing going far beyond what individuals or small groups can achieve by themselves.

It is no coincidence that the first advanced civilizations developed where communities with large populations in arid regions were compelled to improve the natural fertility of their fields through irrigation–around the Euphrates, Tigris, Nile, Indus, and Yangtze rivers.

Irrigation helped these civilizations develop, but sometimes it was also the unexpected cause of their decline. Over the decades, mud, debris and sediment clogged their intricate canal systems and reduced productivity. Because of intense irrigation and the use of relatively salty water, soils were leached out and became salinized, resulting in lower crop yields that could no longer feed growing populations.

Thousands of years later, at the beginning of the 21st century, this situation has come to a head. In some of the most highly populated regions of the world, people are completely dependent on the irrigation of their agricultural lands. Only in the temperate zones is there enough precipitation to sufficiently water fields during the whole year. More than half of the people in the world live in regions where rains are seasonal or too scarce and unreliable to sustain agricultural production. Without irrigation, their agricultural systems would yield only a fraction of the food they need.

Since the 1960s, many countries, especially in Asia and North Africa, have put tremendous effort into expanding irrigated areas on a large scale. Above all they have concentrated on the construction of dams, canals, and pipelines intended to transfer water to regions lacking this precious resource–sometimes halfway across continents (→ pp.149). In some places such as the Middle East and North Africa, where surface water is scarce, non-renewable groundwater from large depths has been intensively pumped up since the 1970s, effectively mining and diminishing these precious resources for good (→ pp.170).

Irrigated farmland in Asia has more than tripled since the 1960s, and more than doubled in all other regions of the world. Thanks to irrigation, it has been possible to increase both the area of agricultural land and the intensity of cultivation. Farmers now harvest 2 to 3 crops a year instead of only one crop as before. Where people once had to be content with undemanding crops like millet and corn, they can now cultivate more profitable crops like rice, wheat, fruit, and vegetables, which all demand large amounts of water. While only 18 percent of the world's farmland is irrigated, it produces 40 percent of the world's total agricultural output. In most places, the introduction of irrigation coincided with the widespread industrialization of agricultural technology. This adversely affected millions of small and subsistence farmers. Water and grain became more expensive, cultivation intensified, and increasingly fierce competition forced many of those who previously had manually tilled and watered their fields to sell to big landowners and international food corporations. These new landowners in turn consolidated small farms and set up modern, profit-oriented enterprises. As a result of the extensive use of fertilizers and pesticides, high-tech mechanized cultivation, and new kinds of high-yield seed, global agricultural production has almost doubled in the past forty years and the number of people going hungry declined by almost half.

Yet 840 million people, equivalent to more than the total population of Europe, are still chronically undernourished. Some 95 percent of them live in the developing or emerging nations of Asia, Africa, the Middle East, Latin America and the Caribbean.

At the 1996 World Food Summit in Rome, held by the Food and Agriculture Organization (FAO) of the United Nations, delegations from 182 countries adopted a plan of action. This plan aimed to halve the number of people suffering from hunger to 400 million by 2015. It is clear today that this ambitious goal will most likely not be met. By that time, the world's population will have grown by another one billion. To keep up with this population explosion, the agricultural output in developing and emerging nations will have to increase by more than 25 percent over the next ten years, a rate 10 times faster than in the past fifteen to twenty years.

Since the 1980s, the world's population has been growing at a faster pace than agricultural production. The green revolution in developing countries has reached its limits. All over the world, including North America, Europe, and Australia, areas under cultivation can be increased, if at all, only if even bigger and more extensive irrigation systems are installed at huge expense.

The 20th-century vision of an agrarian revolution which would overcome world hunger has not materialized. The finances needed are far beyond the means of developing countries, and solidarity and the willingness to help are apparently too much to expect from richer nations. Some 22 billion dollars a year would have to be invested alone in the development of additional water resources and the expansion of irrigation systems.

The harvest was destroyed by a drought that lasted several months. A peasant in the South Philippines in 2002.
Romeo Gacad/Keystone

Indeed, as things now stand, hunger is due not so much to a shortage of food as to the highly uneven distribution of economic wealth and political power. Lack of money rather than a lack of food is why millions of people go hungry. Wars, corruption, and nepotism in the Third World contribute to the abysmal state of agriculture in many poorer countries. But largely to blame are the economic and trade policies of the most powerful agrarian states, compounding the situation even more.

Agrarian experts widely agree that the security of our food supply is at risk in the long run, not only in the Third World, but everywhere else too. This has many reasons. Agricultural yield, which increased considerably since the 1960s due to irrigation, new methods of cultivation, and improved seed, clearly started diminishing several years ago. New intensified methods of irrigation and cultivation have themselves been one cause. Decades of overuse have leached out the soil, leaving it barren. Dry, crumbly top soil has either washed away or been blown off by storms. In some areas, land has become silted and swamped because excess water has not been effectively drained off. Fields have become salinized, especially in hot drylands where often almost half the water evaporates before it reaches plants. Salinization also poses a threat to farmlands irrigated with non-renewable groundwater, which usually contains higher levels of salt than surface water does. Finally, fertilizers and pesticides pollute freshwater resources to such an extent that in some places water cannot be used for agricultural irrigation, let alone for drinking.

The stealthy degradation of arable land is a malicious and highly unpredictable time bomb. According to agricultural experts, 2 to 7 million hectares of agricultural land are currently lost each year. Almost a fifth of farmland around the world is already moderately or heavily affected, and in Africa, Asia, and Latin America, it is as much as 38 to 74 percent of the land. Scientists at the International Food Policy Research Institute have determined that crop yields on 16 percent of the world's farmlands have noticeably decreased. It is this kind of gradual degradation, going on imperceptibly over decades, that makes the situation so perilous. Once degradation progresses beyond a critical point, enormous efforts are required to recover the land and sometimes it is even lost forever.

The use of fertilizers and pesticides has become one of the most serious problems for modern agriculture. The use of liquid manure and chemical fertilizers has significantly increased global food harvests, and the use of pesticides has significantly decreased crop failures. But at the same time, the nitrates and phosphates in fertilizers, and many highly toxic insecticides and herbicides, have severely damaged farmlands. Not only that—these substances have broadly polluted nature in surrounding areas, affecting streams, rivers, lakes, wetlands, and groundwater, bodies of water almost impossible to cleanse of pollutants.

The use of pesticides has more than tripled from just under one million tons in 1960 to 3.75 million tons in 2000. By 2020, this figure may reach 6.55 million tons. It is anticipated that the use of fertilizers will continue to increase just as rapidly in the coming decades—from about 142 million tons in 2000 to at least 210 million tons. The consequences are staggering. Some 130 billion cubic meters of water polluted with fertilizers seep into groundwater or flow untreated into rivers, lakes and seas every year. Furthermore, wherever intensive animal farming takes place, antibiotics and growth hormones are increasingly detected in water. We are far from knowing exactly what consequences these substances will have in the long run on the health of animals and humans.

In the developed countries of Europe and North America, pollution caused by fertilizers and pesticides has been fairly effectively combated by relatively strict legislation and bans in recent years. Numerous heavily polluted rivers and lakes have indeed partly recovered. But in many developing and emerging countries, the situation continues to deteriorate from year to year. In many African countries, even banned pesticides such as DDT are still being used in huge quantities.

In developing countries, especially in Africa, and even in central and eastern European countries, some 500,000 tons of old, highly toxic pesticides are still largely stored in poorly protected warehouses or out in the open. To leave no residue, these chemicals must be incinerated in special ovens at temperatures higher than 1,400 degrees centigrade. Hardly any African country boasts such a facility, which makes it necessary to transport these hazardous pesticides to North America and Europe as safely as possible—taking them back to where most of them were manufactured in the first place.

According to the FAO, farmland would have to increase by at least 20 percent and water use by at least 14 percent over the next 25 years to meet the growing demand for food. But these are rather unrealistic expectations. It is precisely in developing countries that water is already becoming more and more scarce as both industrialization and the growth of urban centers (→ pp. 227) are demanding an ever-growing proportion of the available resources.

Advocates and opponents of the present development alike are rightly calling for a new, "blue" revolution. The efficiency of irrigation systems, in particular, is meant to improve along the lines of "more crop per drop." Experts have high expectations of drip or microirrigation technology, which supplies water to plant roots in small, precisely measured amounts, thereby greatly reducing the loss of water through evaporation and seepage (→ pp. 205).

Experts disagree however on the use of genetically-modified seed. The FAO, World Bank, and highly-industrialized western countries, backed by agro-chemical corporations, firmly believe that crop yield could be raised a lot by using genetically modified seed. But reliable prognoses are not available. Critics point out that the effects of genetically modified plants on the rest of the plant and animal world haven't been researched enough and that they could lead to the extinction of numerous crop varieties that are well adapted to local conditions, making them indispensable for agriculture.

Contrary to the optimistic predictions made by advocates of genetically modified plants, critics expect the food supply problems of the poorest countries to worsen. Those most in need, small farmers and those who are self-sufficient but without funds, can't afford the new seed they would be forced to buy from European and North American seed companies. Many would come under additional pressure to give up their farms and sell to big landowners and corporations. Instead of remaining self-sufficient, they would have to buy their food from a market that grows mostly for exports and concentrates on the production of more profitable grains, fruits, and vegetables. These foods are too expensive for poor populations.

Development experts call for locally adapted solutions to help the poorest members of the population provide for themselves. Instead of depending on expensive seed and the implementation of costly, large-scale technology, agricultural methods should be adapted to the local climate and existing social structures. Unlike any large-scale technology strategies geared towards increasing profits to the neglect of all other factors, this approach is aimed at using the land in a sustainable way. Productivity can be maintained only if soil and water are used carefully and sparingly.

Advocates of sustainable agricultural policies have proposed a number of soft measures. Traditional methods for storing water should be encouraged, like harvesting rainwater and dew. Native plant varieties needing little water or fertilizer and no pesticides should be grown again. Improving education and setting up cooperatives and organizations to monitor the fair distribution of limited resources would allow small farmers to become self-sufficient, at least for essentials. In short, the cultivation of staple foods for local markets should be promoted, not the production of fodder cereal for livestock, or fruits and vegetables for export.

These demands are not new. The 1996 World Food Summit in Rome adopted a plan of action which outlined the principles of such a reform. This plan of action states that agricultural and nutritional policies are needed that will permit farmers to increase their production, and encourage conservation and sustainable management of natural resources. The plan goes on to say that creating appropriate social and political framework conditions is necessary for safeguarding fair agricultural policies and ownership structures, reliable access to water, seed and fertilizers, and easier access to affordable loans. Furthermore, provisions should be made for improving educational opportunities and encouraging new, ecologically sound technologies. Last but not least, structures should be set up to enable developing countries to store, process, transport and market crops themselves.

Development organizations also call for an international agricultural policy that doesn't undermine these goals, but favors them. The World Bank and the International Monetary Fund, the most important international financial institutions, are not willing however to take the appropriate steps, nor are rich industrialized nations who keep protecting the interests of their own national agriculture industries and construction and water companies (→ pp. 498).

But even if all reasonable measures could be put into practice, two main problems would remain unsolved. Nearly all countries in the Middle East and North Africa lack the natural water resources they need to grow enough food for their populations. Even when optimal methods of cultivation are used, crops need at least three cubic meters of water to yield one person's daily ration of food, which amounts to a minimum of 1,100 cubic meters of water per person per year. In Morocco and Egypt, almost that water quantity, 990 and 880 cubic meters of water respectively, are actually available. But countries like Algeria (594 m^3), Israel (450 m^3), Tunisia (444 m^3), Saudi Arabia (160 m^3), and Libya (100 m^3) have nowhere near the required quantity. Expanding populations will cause the quantity of water resources available per person to further decrease by at least a third. Many countries in the Middle East and North Africa will depend on huge food imports, even in the distant future, and thus the poorest countries will continue to depend on humanitarian aid for survival.

The second forecast is even more alarming. Water experts estimate that within two decades, expanding populations will push global water consumption beyond accessible annual global precipitation–beyond the total amount of renewable water resources. The only way of counterbalancing this enormous water deficit would be to exploit nature even more by destroying wetlands, bogs, moors, water meadows, natural grasslands, recreational areas, and forest regions. Or by diverting even more water from rivers. This could mean that all the water from major rivers would be utilized, effectively draining the Ganges, Brahmaputra, Indus, Mekong, and Yangtze rivers in Asia, as well as the Nile in Africa, leaving their deltas barren and dry. Even in wetter regions, this would have an enormously negative effect on the ecological quality of the natural environment, possibly leading to irreversible damage to whole continental and oceanic regions.

In arid regions, one single crop failure caused by lack of rain may lead to severe food crises: food distribution in Gabi, Niger, 2005. Geert van Kesteren/Magnum Photos

Trade and global markets determine the conditions of production. Tomato farmers between Almeria and Nijar in southern Spain. Photo taken in 2004. Stuart Franklin/Magnum Photos

In the international agri-business, water, as a production factor, is now totally subject to the logic of economics. Wholesale agricultural market in Rungis, France. Jean Gaumy/Magnum Photos

The dictate of prices entails industrial cultivation methods such as intensive use of chemicals and irrigation. Supermarket in Neuilly-sur-Seine, Paris. Patrick Zachmann/Magnum Photos

Agricultural policies gone bad: The destruction of surplus oranges, the result of overproduction. Paterno, Sicily, 1984. Ferdinando Scianna/Magnum Photos

Der Anteil der Landwirtschaft am Wasserverbrauch

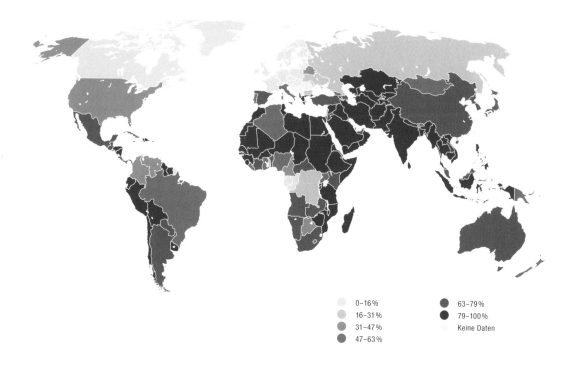

- 0–16%
- 16–31%
- 31–47%
- 47–63%
- 63–79%
- 79–100%
- Keine Daten

Agrarökozonen
Welche Getreideart gedeiht in welcher Region am besten?

- Weizen, Gerste, Roggen
- Reis
- Mais
- Sorghum
- Hirse, Futterhirse
- Nicht geeignet

Bewässerte Fläche als Prozent der Gesamtanbaufläche (1961/1997)

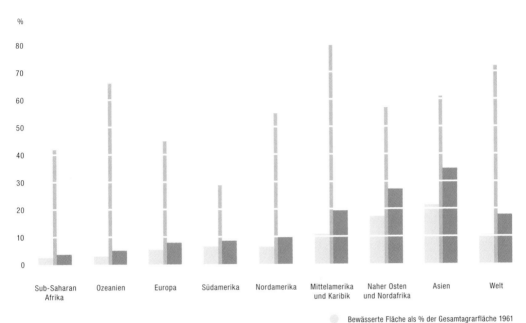

Bewässerte Fläche als % der Gesamtagrarfläche 1961
Bewässerte Fläche als % der Gesamtagrarfläche 1997
Zuwachs der bewässerten Fläche in %

Ländervergleiche nach verschiedenen Kriterien

	Unterernährte in % der Gesamtbevölkerung	Bewässerte Fläche in % der Gesamtanbaufläche	Wasserentnahme für Landwirtschaftzwecke in % der erneuerbaren Wasserressourcen
Ägypten	4	100	93
Brasilien	10	4	0.4
China	9	39	14
Deutschland	0	4	6
Indien	23	34	31
Iran	5	40	49
Israel	0	45	78
Libyen	0	22	854
Simbabwe	39	3	11
Spanien	0	20	22
Thailand	21	26	19
USA	0	12	7
Usbekistan	4	88	108
Vereinigte Arabische Emirate	0	56	1021

During the p
there has be
increase in
irrigated crc
so, almost o
people suffe
malnutrition

ast **50 years,**

en a **threefold**

artificially

o land. Even

e billion

r from chronic

In Egypt, reliable irrigation safeguards several vegetable harvests per year. The Nile Delta. Photo taken in 1998.
Ian Berry/Magnum Photos

Nowadays, water can be pumped out of the Nile and into irrigation ditches whenever necessary. Before the Nile was regulated by the large Assuan dam in 1971, the peasants planted their crops in harmony with the natural high waters of the river. Khaled El Fiqi/Keystone/EPA

Dams and irrigation systems—temples of modernity

At the inauguration of the Bhakra Dam, designed by Swiss architect Le Corbusier, India's prime minister Jawaharlal Nehru described it as one of the first "temples of modern India." India built over 4,000 such temples in the forty years that followed.

The opening of the Bhakra Dam on October 22, 1963 signaled the beginning of India's green revolution, the most ambitious attempt ever to eradicate hunger in India. Many developing countries in Asia and Africa followed suit, their governments proclaiming a green revolution too. Within a few decades, traditional agriculture in the hands of small farmers was supposed to be catapulted into the 20th century by means of new high-yield crops, fertilizers, pesticides, modern mechanized methods of cultivation, and extensive irrigation systems. At the same time, rapid industrialization was supposed to greatly improve income and the quality of life, especially in cities. Developing countries in Asia in particular decided to construct major dams, hydroelectric power stations, and extensive networks of canals, pipelines, and irrigation systems to meet horrendously growing demands for water and energy.

Since the 1950s, the number of major dams around the world has increased more than sevenfold, from 5,750 to over 47,000. According to the World Commission on Dams, more than half of them (56 percent) are in China and India alone, the world's two most populated countries. Several thousand more are currently being planned or built. China leads the list with 22,000 large dams, over 46 percent of all dams worldwide. Most Chinese dams serve both agricultural and energy purposes (→ pp. 320). Chinese hydroelectric power stations have an installed capacity of 74 million kilowatts, covering more than 18 percent of domestic power consumption. China aims to increase this capacity to over 100 million kilowatts when the Three Gorges Dam begins operating in 2009.

The main dam of the Three Gorges project is 185 meters high and 2,300 meters wide. It will turn the waters of the Yangtze into a reservoir 640 kilometers long. Like many other dams, this project is highly controversial. Roughly 1.9 million people will be forced to resettle. It is still not at all clear what ecological effect the Three Gorges Dam will have on the hydrologic balance of the lower reaches of the Yangtze.

Numerous canal systems, devised to channel water for industry and agriculture over hundreds or thousands of kilometers to arid regions, are similarly gigantic. Thanks to these intricate water networks, the area of irrigated farmland around the world has more than doubled since the 1960s. Some 17 to 18 percent of all farmland today is irrigated, supplying the world with more than 40 percent of its food.

India, home to the most extensive irrigation system in the world, has even bigger aspirations. President Abdul Kalam in August 2003 gave the green light for one of the world's largest infrastructure projects. Ten of India's largest rivers and twenty-seven other waterways are to be linked to form a huge water-distribution complex. In the west, a canal crossing the deserts of Rajasthan province will link the Narmada River with the Tapi and Yamuna Rivers. In the east, the waters of the immense Brahmaputra River will join the waters of the Ganges and other rivers. Water is to be channeled across a system of canals 9,600 kilometers long, involving 32 middle-sized and larger dams and numerous pumping stations, even taking it as high as the Deccan plateau. The plan is for some 173 billion cubic meters of water from the north, mostly from Himalayan rivers, to be transferred each year to the dry regions of eastern and southern India.

Planners hope that this will bring water to an additional 35 million hectares of land that are now barely fertile. The River Linking Project is supposed to be completed within ten years at an estimated cost of $112 billion, double the cost of all irrigation systems India has built during the past fifty years (→ p. 160).

The Chinese Nanshui Beidiao (meaning "water from the south for the north") project is even more ambitious. It dates back to plans laid out by Mao Zedong in the 1960s. This is the largest network of canals ever designed. Three independent canals would divert 48 billion cubic meters of water a year from the Yangtze to the densely populated industrial and urban centers of the north (→ p. 160).

A western canal is to feed 40 percent of the Yangtze's headstreams across the 4,000-meter-high Tibetan plateau into the Yellow River. A central canal is supposed to transfer 40 percent of the Han River across 1,250 kilometers to the Beijing area in its first phase of construction. In its final phase of construction, this canal would also transfer water from the Three Gorges Dam to the Beijing area. Finally, a more than 1,000-kilometer-long eastern canal will transfer water from the lower reaches of the Yangtze through 13 sewage-treatment plants and 30 pumping stations to the Tianjin region. This $60 billion project will displace more than two million people from over 100 villages and towns, and destroy 44,000 hectares of farmland. It will also fundamentally change the lives and working conditions of more than a hundred million people living on the lower reaches and deltas of the diverted rivers.

Not only developing Asian countries, but highly industrialized countries too, especially the United States (2003: 6,500 larger dams), Japan (2003: 2,600), and Spain (2003: 1,200) rely on large-scale dams and extensive irrigation systems. Water is plentiful in northern Spain, but the south and southeast are relatively dry. Since the 1960s, Spain has pursued a questionable water-distribution policy, enabling one of the driest regions in Europe to produce immense amounts of vegetables, citrus, and other fruit crops. To make matters worse, large tourist resorts along the Costa del Sol use increasingly more water. Consumption of water in the south of Spain has gone up 13 percent each year, making these regions completely dependent on water resources from the north. The Spanish parliament in 2001 approved a new project, the National Hydrological Plan, which foresaw building or enlarging 114 dams and tapping as much as 1 billion cubic meters of water from the delta of the Ebro, Spain's largest river, transferring water to the south in a canal system more than 900 kilometers long.

Although several hundred thousand Spaniards demonstrated against this plan and the European Union Commission refused to subsidize it after a critical assessment, the Spanish government saw no reason to give up this 5.4 billion-euro project. In the summer of 2004, the newly elected social democratic government finally put it to an end. Instead of channeling the Ebro, Spain will build 15 large seawater desalination plants along its coasts.

But this will not be enough to solve Spain's future water problems. The Spanish ministry of the environment estimates that by 2050 the average annual temperature will increase by 2.5 degrees centigrade while precipitation will drop by 10 percent and soil moisture by 30 percent. The demand for water is expected to go up much faster than it has in the past. Yet 31 percent of Spain's total area is already threatened by soil degradation to the point of desertification. For Spain to manage its water problems in coming decades, it will have to vastly improve water efficiency and learn to use and reuse water several times over.

Meanwhile, the enthusiasm for dams has ebbed. Large-scale projects, heavily subsidized until now by western industrial nations and the World Bank—usually for self-serving reasons—are being viewed more and more critically. Planners and engineers doubt that dams can actually deliver the performance expected of them. Economists also question whether their huge costs will ever amortize. Mass demonstrations by the people affected, and criticism from ecologists and environmental organizations, have drawn attention to the serious social and long-term ecological effects of these gigantic projects. The World Commission on Dams (WCD), founded in 1997 with support from the World Bank, was called together so that advocates and opponents, representatives of affected populations, NGOs, environmentalists, the private sector, governments, and research institutes could meet and jointly take stock of experience gained so far.

The WCD's final report, released in November 2000, did acknowledge the positive aspects of dams, but came up with a predominantly negative assessment. From an economic point of view alone, dams in many areas had failed to meet expectations. The commission found that the net result for major dams was more or less positive only for power production and flood control. Regarding irrigation, the report stated that: "Large irrigation dams have typically fallen short of physical targets, failed to recover their costs, and been less profitable in economic terms than expected." Furthermore, large dams built for municipal and industrial water supply "have generally fallen short of intended targets." The report found that projects had often greatly exceeded their estimated costs and that investments needed for safety and maintenance were much higher than expected when dams aged. Additionally, sedimentation and the consequent long-term loss of storage is a serious concern globally.

The commission's assessment of the social and ecological impact of dams was also negative. Between 40 to 80 million people have been driven away from their villages in the last fifty years. The Indian Narmada project alone flooded 245 villages and forced 200,000 people to move. Furthermore: "It is not possible to mitigate many of the impacts of reservoir creation on terrestrial ecosystems and biodiversity, and efforts to 'rescue' wildlife have met with little sustainable success." The commission's conclusion on the ecological impacts of dams is worded diplomatically: "On balance, the ecosystem impacts are more negative than positive and they have led, in many cases, to significant and irreversible loss of species and ecosystems."

Experts particularly criticize the fact that the debt service for large-scale projects forces poor countries to neglect the maintenance of existing irrigation systems and dams. The result is that at least 40 percent of the water seeps away before reaching the fields.

The long-term ecological damage done to natural surroundings by these massive interventions can be neither foreseen nor financially assessed. Some of the world's biggest rivers—the Nile, Ganges, Yellow River, Yangtze, and Rio Grande—dry to a trickle in the summer months and never reach their estuaries. The widespread destruction of wetlands and forests falling victim to the construction of dams has resulted in the extinction of many plant and animal species. The loss of forests and wetlands, along with evaporation and changing water temperatures in lakes and seas, influences local climates. Recent studies indicate these factors may even be responsible for climate change on regional and global levels.

The preference for such large-scale projects can hardly be explained in rational and strictly factual arguments. Dams are frequently emotionally charged political symbols. They are temples of modernity, cathedrals of belief in technology, progress, and future prosperity. It is no coincidence that the United States secretary of the interior, Harold Ickes, declared the Hoover Dam at its inauguration in 1935 to be a victory of man over nature. The dam was meant to restore the self-confidence of the American people in their own determination and strength after the 1929 collapse of the economy (→ p. 161).

The Hoover Dam, which holds back the waters of the Colorado River, creating a reservoir 160 kilometers long, is of course highly beneficial economically. It supplies water to mining centers and newly laid-out agricultural lands in the arid areas of Arizona and Nevada. It also provides some of California's densely populated metropolitan areas with energy and drinking water. At the same time it is supposed to be a triumphant monument to the victory of mankind over nature. Nothing makes this clearer than its most absurd product, Las Vegas, a surreal city built in the middle of the desert. It is the essence of the American Dream, an artificial paradise built in defiance of nature—promising luck, riches, and entertainment. It is also the most needless and senseless waste of water imaginable.

The highly political symbolism of such projects makes it difficult to soberly assess their advantages and disadvantages based on rational criteria. The WCD report carefully and somewhat vaguely stated that: "The World Commission on Dams concluded that the 'end' that any project achieves must be the sustainable improvement of human welfare. This means a significant advance of human development on a basis that is economically viable, socially equitable, and environmentally sustainable."

In spite of this critical evaluation, the World Bank in 2003 clearly reversed its policy and suspended a seven-year moratorium on the funding of major dams. The increasingly pressing need for renewable sources of energy has put hydropower back in the limelight. In coming decades, China, India, and the emerging Asian nations will require a lot more energy for rapid industrialization. They are determined to complete their huge dam projects, and they enjoy the full support of western industrialized nations, whose construction companies hope to profit most from these projects (→ pp. 337).

The straight line as a symbol of control over water: The Californian aqueduct system carries water to irrigate the Central Valley. Constantine Manos/Magnum Photos

The cultivation of opium poppies, Tirili, Afghanistan. Bathgate/laif

A settlement and irrigation project for Tuaregs on the Niger. Ulutuncok/laif

Tajikistan. Marion Nitsch

Qasmiveh, Tyre, Lebanon. Photo taken in 2001. Mohamed Zaatrai/Keystone

Measure of the overutilization of natural waters

Average annual discharge of rivers minus the water used for irrigation: If irrigation consumes more water than the natural water cycle supplies, the water reserves diminish. At the same time, there is a shortage of water to supply the population and ensure an intact river eco-system.

Water deficit in 10^6 m^3/year	Water surplus in 10^6 m^3/year
3,000–2,000	0–500
2,000–1,000	500–2,000
1,000–0	2,000–4,000
	4,000–5,000

Irrigation potential

Technically possible increase in grain production through irrigation

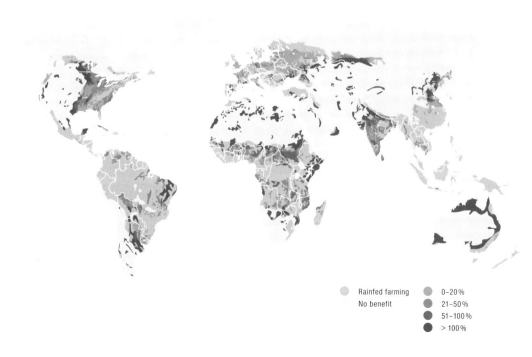

Rainfed farming	0–20%
No benefit	21–50%
	51–100%
	>100%

Percentage of irrigated agricultural area in China (2005)

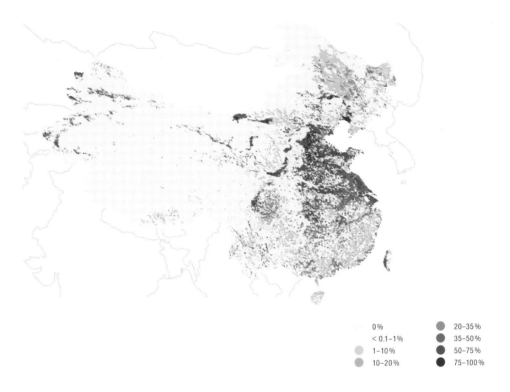

0%		●	20–35%
< 0.1–1%		●	35–50%
●	1–10%	●	50–75%
●	10–20%	●	75–100%

Percentage of irrigated agricultural area in India (2005)

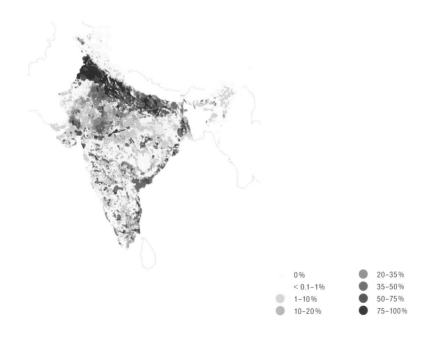

0%		●	20–35%
< 0.1–1%		●	35–50%
●	1–10%	●	50–75%
●	10–20%	●	75–100%

The River Linking Project in India

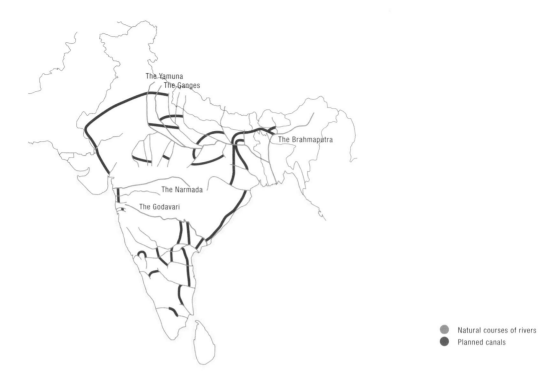

The Yamuna
The Ganges
The Brahmaputra
The Narmada
The Godavari

Natural courses of rivers
Planned canals

The Nanshui Beidiao canal project in China

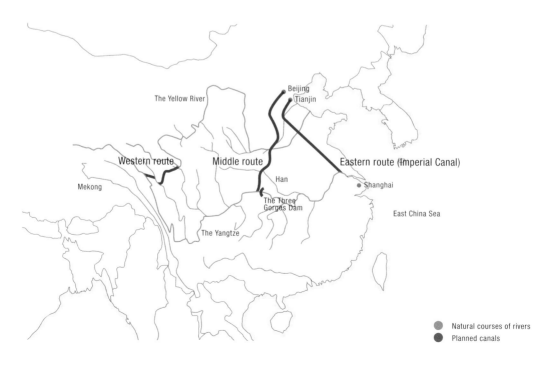

The Yellow River
Beijing
Tianjin
Western route
Middle route
Eastern route (Imperial Canal)
Mekong
Han
Shanghai
The Three
Gorges Dam
East China Sea
The Yangtze

Natural courses of rivers
Planned canals

The Colorado River
Detailed map

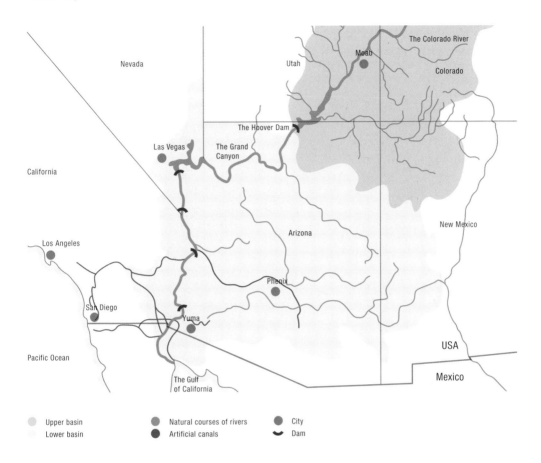

The Colorado River

Nevada

Utah

Moab

Colorado

The Hoover Dam

Las Vegas

The Grand
Canyon

California

Los Angeles

New Mexico

Arizona

Phenix

San Diego

Yuma

USA

Pacific Ocean

Mexico

The Gulf
of California

Upper basin		Natural courses of rivers		City	
Lower basin		Artificial canals		Dam	

Average water discharge in the Grand Canyon before and after the construction of the Hoover Dam

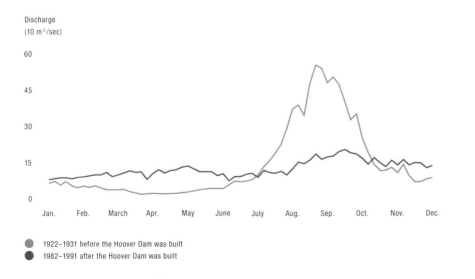

Discharge
(10 m³/sec)

60

45

30

15

0

Jan. Feb. March Apr. May June July Aug. Sep. Oct. Nov. Dec.

1922–1931 before the Hoover Dam was built
1982–1991 after the Hoover Dam was built

The Midwes
the USA is **e**
dent on wat
Ogallala Aqu
continues at
rate, **the wo**
largest grou
reservoir wi
the next thi

Corn Belt of
ntirely depen-
er from the
ifer. If irrigation
the current
ld's third-
ndwater
l dry up within
ty years.

Fields irrigated with sprinkling systems at Kufra in the Libyan desert. A powerful pump in the centre pumps non-replenishable groundwater up to the surface and distributes it with a center pivot sprinkler. Schapowalow/Fotofinder

One of the wonders of the world or a water disaster? The Great Man Made River project in Libya is designed to irrigate dry land by pumping deep-lying groundwater out of the Sahara and to the coast through four-meter-thick pipes. As the water resources are limited, the project will only run for a few years. Luca Zanetti/Lookat

When it ceases to rain, groundwater is often the only hope left. The problem is, however, that groundwater is replenished by rainfall and inevitably dries up when the water used exceeds the amount that the rain and dew can replace. Jebel Aulia refugee camp at Khartoum in Sudan. Abd Raouf/Keystone

Groundwater
is running dry

In the Indian village of Manerajree, as described by scientist Vandana Shiva in *Blue Gold: The Fight to Stop the Corporate Theft of the World's Water*, sugar-cane farmers in November 1981 built a small irrigation system. For just under $14,000, they drilled three 60-meter holes in the ground to extract 50,000 liters of water a day for their fields using new motor pumps. At first the three new wells provided them with exactly the amount of water they had expected. But little more than a year later, the wells dried up—their groundwater supply had been exhausted. Because the groundwater table had sunk, some 200 shallow wells in the surroundings also dried up. These wells had been a modest but steady source of water for centuries. Since then, sugar-cane plantations have had to be supplied with water brought in tanker trucks.

What happened in Manerajree in the space of just a few years is in danger of happening during coming decades in many other regions of the world. Aquifers, underground layers of water-bearing permeable rock that allow the passage of groundwater, are drying out. Even experts at the International Water Management Institute, certainly not otherwise known for propagating worst-case scenarios, consider this to be one of the most serious problems facing global water management.

Ever since deep drilling and high-capacity pumps have enabled us to obtain water from deeper and deeper sources, global consumption of water has soared. Statistically this is not a problem. The world's groundwater reserves are estimated to be 10.5 million cubic kilometers. This is almost 100 times more than the total freshwater reserves in circulation on the surface of our planet.

Roughly one-fourth of the world's population uses groundwater for its supply of drinking water. Many countries in the Middle East and Asia, among them China and India, obtain at least half of their drinking water from groundwater reserves. This is also true of the Netherlands, Denmark, and Barbados. These sources are mostly near the surface and are therefore rapidly replenished by precipitation.

Fossil groundwater deposits are however a different story. They are among Earth's non-renewable resources. Once they are depleted, it can take centuries or millennia for them to be replenished, if at all.

This makes it all the more astonishing that Earth's last abundant freshwater reserves are being used so wastefully and on such a large scale for irrigation. Since the 1960s, India alone has installed almost 20 million motorized high-capacity pumps, half of which have been installed since 1990. Each year, they pump approximately 244 cubic kilometers of water to the surface, twice the volume that seeps back into the groundwater over the same period of time. The consequences are already severe. The groundwater table in several Indian states has fallen by more than 30 meters since the 1970s, and some major aquifers have been totally depleted.

In the Indian state of Tamil Nadu, the country's "rice bowl," the green revolution has been responsible for emptying out the Cauvery River, once 300 meters wide. Today it is sometimes completely dry. The ground-water table has dropped by 300 to 400 meters in many parts of this region. Roughly 1 million small farmers have lost their means of subsistence because they can't afford high-capacity pumps and aren't connected to an irrigation system.

Eleven countries holding almost half of the world's population, among them China, India, Pakistan, the United States, Israel, Egypt, Libya, and Algeria, already have a negative water balance, in other words, they use up more groundwater than nature returns (→ p. 180). But national water balances only tell us half the story since groundwater problems occur on a local or regional scale. It's when water dries up in an agriculturally important area like a "breadbasket" that a whole country or even inter-national markets are affected. Farms and plantations must then readapt and learn to make do once again with water from precipitation and rivers.

The Ogallala Aquifer in the United States is the most striking example. Although quite shallow, it is the world's third-largest groundwater reservoir, extending over eight states from Texas in the south to South Dakota in the north—an area approximately the size of France. The Ogallala Aquifer supplies water to approximately one-fifth of the irrigated farmland in the U.S. (→ p. 180)

Starting in the 1950s, the American agricultural industry in this region set up huge farms for cultivating cotton, wheat, and alfalfa. Within just a few years, farmers began operating 17,000 high-capacity pumps and installing irrigation pipelines and sprinkler systems that extended over vast areas. Plants to process cotton and wheat for export followed. An entire indus-trial infrastructure was created for stockyards holding up to 20,000 head of cattle, including large-scale slaughterhouses, and meat processing and packing plants.

By the end of the 1970s, just twenty years later, the groundwater table had already sunk by several meters in many places. Natural wetlands dried up. Grazing land and forests beyond the irrigated areas increasingly suffered from water scarcity.

Farms started to directly feel the effects of this exploitation. The exces-sive use of fertilizers and pesticides depleted and contaminated the soil, and the extremely high evaporation rate of sprinkler systems caused fields to become salinized much faster than had been expected. At the same time, falling grain prices on global markets prevented farmers from being able to make additional investments to compensate for stagnating productivity.

Since then, the cultivated area irrigated by the Ogallala Aquifer has been reduced by more than 20 percent, and in some places by as much as 60 percent. Many farmers have been forced to return to the conventional method of rainfeeding their fields. Instead of producing fodder grain and cotton, they now cultivate crops like millet, less demanding but lower in yield. The number of livestock had to be drastically reduced in many stockyards, and others had to close down. The processing industry was forced to cut back operations as well, depriving tens of thousands of farmers and industrial workers of their jobs. If the Ogallala Aquifer, the third largest in the world, continues to be exploited at the same rate as in the past decades, experts predict that it will stop yielding water in twenty to thirty years.

New groundwater-mining technology has sometimes inspired completely utopian projects in arid regions of the Third World. In the 1970s, Libyan leader Muammar al-Qaddafi tried to make his dream of a green paradise in the middle of the Kufra Desert come true. The Nubian Sandstone Aquifer, 2,000 meters underground, is the world's largest freshwater reservoir. Qaddafi intended to use its water to irrigate 50,000 hectares of desert, turning it into the world's most spectacular breadbasket. Ten years later, when this bold experiment literally filled up with sand, Qaddafi launched an even more spectacular project, the Great Man-Made River, which is supposed to transfer two million cubic meters of water from the Nubian Sandstone Aquifer through a 3,000-kilometer pipeline to the coastal regions of Benghazi and Tripoli. Should this project ever be completed, Qaddafi believes it will be regarded as the eighth wonder of the world.

The geological and ecological consequences of the exploitation of fossil groundwater reserves in the region are still unclear. In the area around Sana'a, the capital of Yemen, the groundwater table falls by about 3 meters each year. This is because four times more water is pumped to the surface than is replenished by precipitation. In the region around Beijing, where the groundwater table drops by 1 to 3 meters annually, the ground itself is now sinking about 10 centimeters each year. As water is drawn off, the depleted aquifers dry out and cave in. In some of Mexico City's neighborhoods, the ground sinks by as much as 30 centimeters each year. The same can be observed in Bangkok, Manila, Houston, and many other cities. Half of a suburb in Baytown, a harbor community in Texas, sank into the sea with more than 200 houses because the petrochemical industry had drawn off far too much water.

The impact of sinking groundwater tables is particularly serious in coastal regions. Once the groundwater table sinks below sea level, seawater seeps into the drained space and salinates the remaining water. Concentrations of salt in coastal aquifers in Gujarat in India, on the east coast of Spain, in Israel, and in Florida, are already so high that water can't be used for drinking or for irrigation.

All hopes have been dashed that the exploitation of aquifers could relieve water shortages in some parts of the world for a long time. In reality, the water effectively available underground has been much less than expected. Unless the consumption of groundwater reserves is drastically restricted to the natural replenishment rate, aquifers will not last for centuries, but dry out in decades, even years.

At this point we can't assess or even estimate the ensuing ecological, economic, or social consequences, although one thing is clear.
In the long run, mankind will have no other choice but to make do with the approximately 110,000 cubic kilometers of rain which nature lets fall on Earth each year (→ p. 64).

Ten times the amount of water is required to produce meat as to cultivate vegetables.
A cattle farm in San Lucas, California. Marcio José Sanchez/AP Photo

In the Corn Belt in the Midwest of the USA, the groundwater levels are falling because farmers are pumping up far more water than the rain can replace. Ever more acreage has to be abandoned, while the countryside is regaining its original steppe character. Ilkka Uimonen/Magnum Photos

300 grams of meat, 3,000 liters of water. Martin Parr/Magnum Photos

Groundwater use

Groundwater use per person per annum (1998 or the latest available data)

0–100 m³	No data
101–250 m³	Use exceeds annual
251–500 m³	formation
501–900 m³	

Detailed map of the Ogallala Aquifer

Groundwater resouces

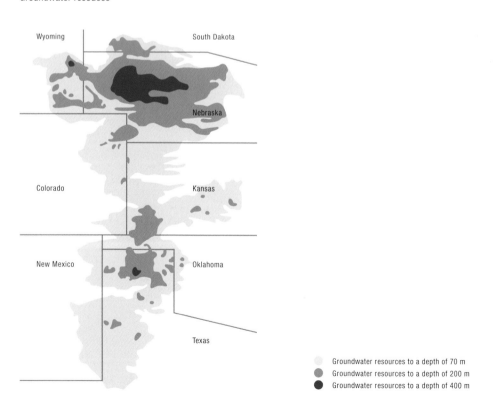

Groundwater resources to a depth of 70 m
Groundwater resources to a depth of 200 m
Groundwater resources to a depth of 400 m

Water consumption per kg produced
Example: irrigation agriculture in California

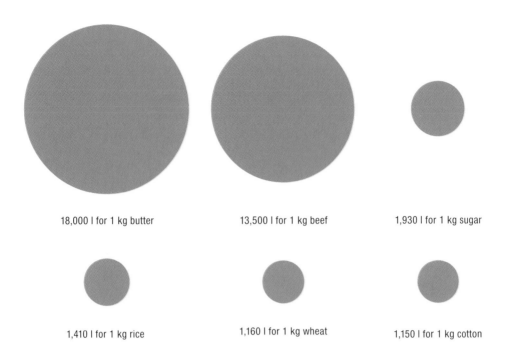

18,000 l for 1 kg butter 13,500 l for 1 kg beef 1,930 l for 1 kg sugar

1,410 l for 1 kg rice 1,160 l for 1 kg wheat 1,150 l for 1 kg cotton

Irrigation methods
The quantities of water required for one month's irrigation

Field crop	Ridge-and-furrow irrigation in m³/ha	Drip irrigation in m³/ha
Sunflowers	2,380	1,980
Wheat	1,850	–
Corn	1,790	950
Field beans	1,240	810
Tomatoes	1,190	950
Citrus fruits	790	400
Grapes	500	380

The drier and the hotter the climate, the greater the difference between ridge-and-furrow and drip irrigation

Due to **wror**
sive use of v
of the worl
acreage is n
resulting in g
ished harves
over **1 milli**
rendered inf

g and exces-
ater, one third
's irrigated
ow salinated,
reatly dimin-
ts. Each year,
n hectares are
ertile as a result.

The white gold is devouring the blue: The cultivation of cotton is one of the most wasteful uses of water. Cotton harvest in Mississippi. Hiroji Kubota/Magnum Photos

Excessive irrigation and poor drainage cause waterlogging. Saline groundwater then rises to the surface, where it evaporates, depositing a centimeter-thick crust of salt. Al Quabla near Bagdad, Iraq.

Markus Metzel/Das Fotoarchiv

Irrigation, soil, and salt

Agricultural experts estimate that every year more than one million hectares of the world's farmlands are irreversibly destroyed by salination. High concentrations of salt have caused varying levels of damage to almost a third of all irrigated fields. The developing countries of Asia and Africa are the regions most affected.

Soil salination is a complex problem. The salinity of soils depends on many factors that can't be fully controlled or regulated. Everywhere in nature, water dissolves miniscule amounts of mineral salts contained in rock and earth. The natural composition of salts varies, depending on local geological and hydrological conditions. The longer water remains in the earth and the more soluble the minerals it encounters underground, the higher its salinity. This is why fossil groundwater stored in the earth for hundreds or thousands of years usually has a much higher concentration of salt than rapidly circulating surface water.

Evaporation also plays an important role in soil salination. The more water evaporates, the more the salt content in the remaining water increases. Soil in hot and arid regions therefore becomes salty much faster than in temperate zones. Large dams, extensive open irrigation ditches, and methods of cultivation that use excessive amounts of water, such as the flooding of rice paddies or the constant running of sprinklers to water cotton and grain fields, result in high rates of evaporation and even greater salination in irrigated fields.

Added to that are huge amounts of salt from various kinds of fertilizers and sewage sludge, and from household and industrial wastewater released untreated into streams, rivers, lakes, and irrigation canals, from where they reach groundwater again.

Wherever water is used over and over again for industrial or irrigation purposes—in the Colorado River it is reused up to eighteen times—the salt content of water steadily rises and salination severely affects the lower reaches of rivers and the wetlands of estuaries.

In less than thirty years, at least half of the cotton fields in Turkmenistan have turned into salty marshlands. High concentrations of salt have caused crop yields to significantly drop in 25 to 30 percent of irrigated fields in the United States.

Cropping patterns and especially drainage can mainly regulate soil salinity. Plants absorb only certain amounts of specific nutrient compounds. If water doesn't contain enough of an essential nutrient salt, plants become stunted or do not grow at all. If the concentration of salt is too high, they wither and die. Unless excessive salt is flushed away by rain or washed out through a good drainage network, it steadily accumulates over decades in the topsoil. Damage becomes irreversible once the salt concentration reaches a certain critical level.

Above all, non-existent or inadequate drainage of cultivated soils is responsible for excessive salination. This is particularly true of overirrigated land where deeper and deeper subsurface layers of soil are soaked with water. When this water finally makes contact with the groundwater underneath, the soil becomes waterlogged and acts like blotting paper. To the same degree that water on the surface evaporates, capillary action causes the (usually much more highly salinated) groundwater to rise back to the surface. Within a few years, fields can become so salty that they are covered with crusts of salt and rendered forever infertile.

Evaporation and salination are particularly devastating when rivers and lakes are subject to other environmentally damaging factors. One of the world's biggest environmental tragedies has been going on for fifty years around the Aral Sea in the border area between Kazakhstan and Uzbekistan. Once the world's fourth-largest lake and in the 1960s more than one-and-a-half times the size of Switzerland, the Aral Sea has since shrunk in half, the volume of its waters reduced by three-fourths (→ p. 196).

It all began with another grand plan. Russian czars started extensively irrigating the steppes to grow cotton in their central Asian colonies. By the end of the 19th century, 2.5 million hectares were being irrigated. By 1950, this figure had increased to 4.7 million and by 1990 it had reached 7.9 million hectares. The area of cotton fields has increased threefold since the 1950s. The area for growing rice, which has been forcefully expanded since the 1960s, has increased sixfold. Added to that, rice needs three times as much water.

The construction of irrigation systems so reduced the flow of the Amu Darya and Syr Darya River feeding into the Aral Sea that the sea is now literally drying up. The Kara Kum Canal, which diverts water from the Amu Darya River to the steppes of Turkmenistan along a 1,500-kilometer long canal, is alone responsible for reducing by some 40 percent the flow of water that reaches the Aral Sea. Long stretches of the canal system have not been sealed with concrete, so much of the water seeps into the ground before ever reaching the agricultural fields of Uzbekistan, Kazakhstan, and Turkmenistan. Experts estimate that up to 60 percent of the diverted water seeps away or evaporates without being used. The small trickle of water that still feeds into the Aral Sea is so heavily contaminated with fertilizers and pesticides that most of this former freshwater lake is now biologically dead. Due to evaporation and the lack of freshwater input, the Aral Sea is now two-and-a-half times saltier than the oceans.

The fisheries industry was an important economic factor in the region, with the lake yielding 44,000 tons of fish each year. Fishery was terminated in 1992 and about 60,000 fishermen and fish processing workers lost their jobs. Some 500,000 hectares of land along the tributary rivers and their deltas have been destroyed because of salination and toxic contamination. Harbor towns and seaside villages now lie up to 100 kilometers from the shore. The vast newly exposed plains of the former seabed extend over several thousand square kilometers, a hostile landscape of highly contaminated silt encrusted with salt.

The damage is huge and can hardly be remedied. Some 1.5 million people are directly affected and another 2.3 million at least indirectly affected by this disaster. The incidence of anemia, cancer and tuberculosis has markedly risen. Furthermore, the Aral Sea once acted as a heat reservoir, stabilizing the region's climate, but since the lake's slow disappearance the climate has noticeably changed. Precipitation has decreased, winters are colder, and summers hotter. Winds swirl up as much as 100 million tons of dust each year from the newly exposed former seabed, carrying with them salt, herbicides, and pesticides, and spreading contamination over the entire region.

Central Asia's second largest closed lake, Lake Balkhash, is threatened by a similar fate. Like the Aral Sea, it has no outlet and its water level is regulated only by evaporation and the balance of its tributaries. Just seventy years ago, this 620-kilometer-long lake in the desert steppes of Kazakhstan was surrounded by unspoiled nature. In the 1930s, large copper and iron ore deposits were discovered and a smelting works employing 15,000 workers was built. Zinc levels in the water are now 21 times above officially set limits, and chromium levels 13 times above.

Like the Aral Sea, Lake Balkhash is in danger of being cut off from the rivers that feed it. The Ili River, originating in the Tianshan mountains of China's Xinjiang region, accounts for 80 percent of Lake Balkhash's inflow. China plans to divert about three-fourths of the Ili's waters to irrigate large agricultural areas in Xinjiang province. Should China carry out this project, Lake Balkhash is doomed to go the way of the Aral Sea within a few decades (→ p. 197).

Sacrificed to cotton exports: The Aral Sea, which was the size of Ireland a mere fifty years ago, has now been reduced to a polluted, saline residue. Ship cemetery on the former lake bed at Muinak, Uzbekistan. Hill/laif

The Aral Sea in the 1960s, with its once rich fishing grounds. NASA

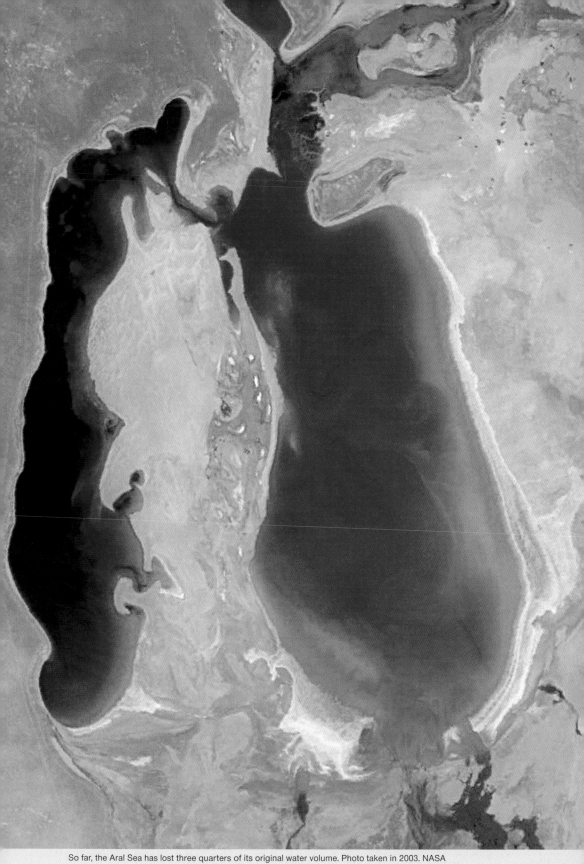

So far, the Aral Sea has lost three quarters of its original water volume. Photo taken in 2003. NASA

Lake Aral (1960/2000)

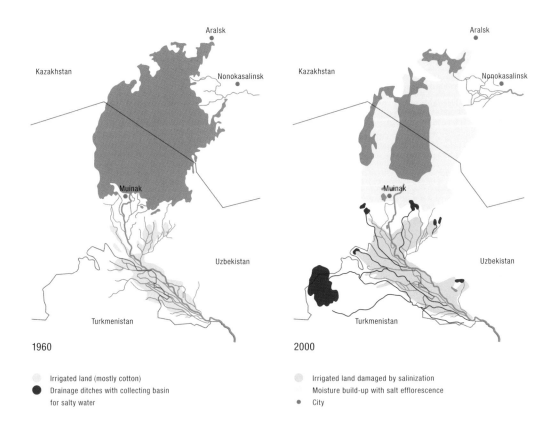

1960

2000

- Irrigated land (mostly cotton)
- Drainage ditches with collecting basin
  for salty water

- Irrigated land damaged by salinization
- Moisture build-up with salt efflorescence
- City

Salinization of soils due to wrong irrigation management

Excessive irrigation results in a rise of groundwater table, sometimes leading to waterlogging of the soil.
Salty groundwater is then drawn to the surface. In hot and dry climates, it evaporates and forms salt deposits on the soil.

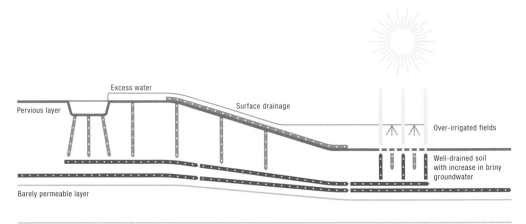

- Groundwater
- High evaporation with salt secretion

Ili-Gebiet, Hauptzufluss des Balchasch-Sees

Entwicklungsvorhaben Chinas im Einzugsbereich des oberen Ili, Nord-Xinjiang

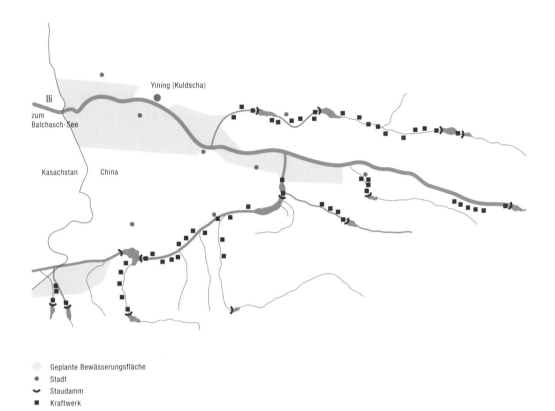

Geplante Bewässerungsfläche
Stadt
Staudamm
Kraftwerk

Prognose für zukünftige Seespiegel des Balchasch-Sees bei vermindertem Zufluss

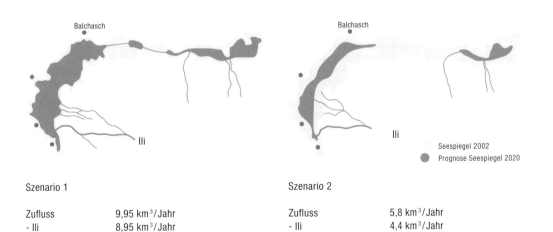

Seespiegel 2002
Prognose Seespiegel 2020

Szenario 1

Zufluss	9,95 km³/Jahr
- Ili	8,95 km³/Jahr
- Östliche Flüsse	1 km³/Jahr
Seespiegel 2020	340 m

Szenario 2

Zufluss	5,8 km³/Jahr
- Ili	4,4 km³/Jahr
- Östliche Flüsse	1,4 km³/Jahr
Seespiegel 2020	337,6 m

On average, [...] of the amou[nt ...] needed to c[...] etable foods [...] required for [...] tion of meat one tenth

nt of water is

ltivate veg-

:uffs than that

:he **produc-**

.

Making the most of every drop: Cultivating tomatoes with an extremely economical trickle irrigation system at Jericho in Jordan. The plastic tunnels ensure that very little water is lost through evaporation. Noel Matoff

In spite of the most imaginative solutions, there are limits to exploiting new water resources: In Chungungo, Chile, local inhabitants use mist-netting to harvest tiny fog droplets in the coastal air. Saussier Gille/Gamma

Even on the virtually rainless Canary Island of Lanzarote, the peasants, drawing on the experience of past generations, patiently cultivate the land. At night, dew is precipitated from the cool sea wind, trickles through the lava sand and is absorbed by the roots of the vines. Playa de Janubio, Lanzarote. Grnelius Maas/Das Fotosarchiv

A rural worker installs drip irrigation tubes to minimize water consumption.
Oase Liwa, United Arab Emirates. Peter Essick/Aurora

Capturing rainwater and using it as efficiently as possible reduces the dependency on rivers and groundwater.
Permaculture Center in Hosar, West Bank. Michael Richter

In search of solutions

The United Nations Food and Agriculture Organization (FAO) estimates that agricultural lands will have to increase by one-fifth in coming years if the UN's millennium development goal for halving extreme poverty and hunger is to be met by 2015. At least 14 percent more water would be needed to accomplish this, even with more efficient methods of irrigation and cultivation.

But it's precisely the large developing and emerging countries that will also need much more water for industry and their rapidly growing populations in coming decades. FAO experts fear that especially in highly populated emerging nations with limited water supplies, even less water will be available for agriculture than at present.

Tapping new water sources isn't going to be enough. This is why development experts advocate more efficient use of existing resources–"more crop per drop." The FAO estimates that water efficiency in developing countries today is only 38 percent. More than half the water is lost by seepage or evaporation before it ever reaches plants. This is mainly due to wasteful methods of irrigation such as the flooding of rice paddies, the use of overhead sprinklers, or furrow irrigation, which conducts generous amounts of water through pipes to individual furrows.

In contrast, drip irrigation, also known as microirrigation, which was developed in Israel in the 1960s, has a water efficiency rate of up to 95 percent. Its principle is simple. Water is applied drop by drop directly to the roots of plants by means of plastic tubes and can be finely measured out. Microirrigation produces 20 or more percent higher yields using 40 percent less water. But this method is currently used on only 1 percent of all irrigated fields (→ pp. 181).

Drip irrigation has its limits and is only suitable for the cultivation of larger individual plants such as grapes, olive and fruit trees, and vegetables. Many small farmers and village cooperatives can't afford effective and durable drip-irrigation systems, even if their fields could be connected to them and they had reliable running water during the growing season (→ p. 364).

But inexpensive drip-water systems are becoming more available. Aid organizations such as the non-profit foundation International Development Enterprises (IDE) have developed a whole range of practical systems that can be easily enlarged, including 20-liter systems for small vegetable gardens to 1,000-liter tanks which can irrigate about 1,000 square meters of land. In the past seven years, IDE has sold 85,000 of these systems to small farmers in India at a reduced price.

Small-scale irrigation systems are also successful in countries with monsoon rains, where traditional methods of water harvesting fall back on collecting and storing seasonal rainwater in barrels, small reservoirs, and ponds. The advantages are obvious—farmers are less dependent on state irrigation systems and the rainwater is free.

In contrast, the great expectations once placed on the development of efficient desalination plants have not materialized. This technology is much too expensive, at least for agriculture, since the desalination process is highly energy-intensive. Depending on the kind of energy used, it costs between one to one-and-a-half dollars to desalt a single cubic meter of water, which is the amount needed to grow a kilogram of grain. More than half of the world's 1,200 large desalination plants are located in very rich countries like Saudi Arabia, Kuwait, Bahrain, Qatar, and the United Arab Emirates, where oil is still very cheap. Among industrialized nations, only the United States operates a larger number of seawater desalination facilities—most notably in California and Florida, where all other cheaper water resources are all but exhausted.

Since the 1990s, agronomists and water experts have been looking at the idea of "virtual" water. This approach is mainly economic and its intention is to improve the efficient use of global water reserves. Virtual water refers to the amount of water needed to produce various kinds of food in different countries. Depending on the climate and the irrigation method used, one kilogram of wheat needs between 1,000 to 4,000 liters of water, a kilogram of rice 1,900 to 3,500 liters and a kilogram of beef at least 15,000 liters (→ p. 181).

Scientists hope that such estimates could be reliable tools for regulating agricultural policy. Studies of different countries show how a switch from fodder cereal to wheat, rice to millet, or cotton to corn would affect a country's water consumption, allowing scientists to calculate how limited water supplies could be used most efficiently from an economic point of view. One such study shows that the cost of supplying water in Saudi Arabia to grow wheat is five times higher than the price of wheat on the world market. Libya too would be better off importing various staple foods instead of producing them at ten times the cost.

The concept of virtual water is controversial, however, among agricultural experts. Critics say these calculation models exclude important factors such as a country's political situation, or the trade barriers between developing countries and powerful industrial and agricultural nations. They fear, in particular, that the concept's pure focus on economics could lead to governments putting economic goals and balanced trade budgets before the basic needs of their populations. They also warn that multinational agriculture corporations could cultivate cash crops for export on large plantations while the local population still went hungry. Last but not least, poor agricultural countries could become even more dependent on the agricultural superpowers of Europe and North America.

How individual national agriculture policies and the global food situation develop is of course much more than just a question of water scarcity and the most efficient use of existing resources. Development policy experts criticize rich industrial and agricultural states for continuing their policy of protectionism, keeping their markets closed to the cheaper agricultural products of less developed nations. Richer nations' policies also prevent poorer countries from engaging in the lucrative business of processing and refining their produce into expensive end products.

International financial organizations apparently adhere to the same agenda. The World Bank, for instance, one of whose chief tasks is to fight poverty, has cut direct agricultural aid to developing nations by a total of 1 billion dollars since the 1990s. It now spends only 1.1 billion dollars each year on long-term rural development projects, as opposed to the 3.9 billion it used to spend. Surprisingly, developing countries themselves have substantially reduced their budgets for rural infrastructure development. In southern Africa, expenditure for rural development has been reduced from 6.2 percent of national budgets to 3.9 percent, in the Middle East and North Africa from 4.1 to 1.1 percent, and in southern Asia from 8.4 to 5.2 percent.

Critics argue that these efficiency-oriented concepts are based solely on economic reasoning. They believe that developing countries' agricultural policies should be primarily concerned with nutritional efficiency, meaning they should make sure that the food is grown which their undernourished populations need to survive. This would presuppose a new order for total global food production. But the economists who currently shape the development policies of international organizations are far from thinking along these lines.

Affluence and thoughtlessness: Alarmingly, the link between production and consumption of food has been severed. Ocho Rios, Jamaica. René Burri/Magnum Photos

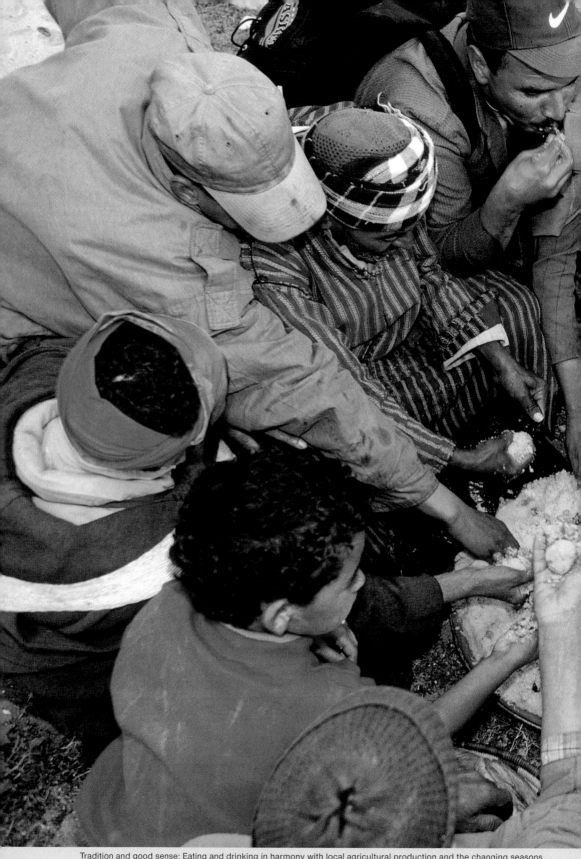

Tradition and good sense: Eating and drinking in harmony with local agricultural production and the changing seasons. Moulay Kertoun near Essaouira, Morocco. Bruno Barbey/Magnum Photos

Water for people

Human beings cannot survive without water for more than three or four days. Like air and food, water is basic for life—no human or animal can do without it.

Access to water for drinking and household purposes is very unevenly distributed in the world. In highly-developed countries, clean water flows from the tap 24 hours a day. Per capita consumption of water is at least 100 liters a day and often much more. But in Africa, Asia, and Latin America, hundreds of thousands of people have to walk miles just to get a few liters, and millions make do with 20 liters or less per day. As if that weren't enough, limited water supplies are not always safe for drinking and sometimes come close to running out for days or weeks on end.

One of the most urgent of modern-day problems is the scarcity of water in cities, particularly in the sprawling slums of rapidly growing megacities. Setting up water utilities that work reliably is an extremely expensive venture and poses an enormous technical challenge. Water in big cities is often more heavily polluted with excreta and harmful chemicals than it is in rural areas. One billion people on Earth have no reliable access to safe drinking water, and two billion people don't have access to the simplest sanitary facilities like latrines, least of all toilets and washing facilities with running water.

Every year, 1.7 million people die from the direct and indirect consequences of poor hygiene and exposure to contaminated water. Roughly 80 percent of all cases of disease in developing countries—hundreds of millions every year—are a direct or indirect result of water pollution and water scarcity.

The international community has not delivered on any of its resolves to overcome water shortage, failing to meet goals even though targets have been set more and more modestly in recent decades.

But there have been successful ventures in cities and rural regions wherever an affected group of people, supported by development and aid organizations, have taken matters into their own hands to mitigate the worst effects of water scarcity.

Living on the brink: When people do not even have enough water to drink, life becomes hell and turns into a battle for survival. In the village of Natwarghad in the Indian state of Gujarat, a single well became the sole source of water, following a prolonged drought in 2003. Amit Dave/Reuters

Urban water shortage: Hundreds of millions of people live in slums and favelas in the megacities of Asia, Africa and Latin America. Most of them have no access to tap water and live in precarious hygienic conditions. The northern favelas of Sao Paulo, Brazil, in 2002. Stuart Franklin/Magnum Photos

1 billion hu[man]

have no ac[cess to]

drinking wat[er]

2 billion hu[man]

live under in[...]

hygienic co[nditions]

han beings

ess to safe

er.

han beings

tolerable

ditions.

The poverty line: The protective fence around a gated middle-class community in northern Buenos Aires directly borders on the slum of La Cava, which is not connected to the public water supply. Photo taken in 2003. Natacha Pisarenko/AP Photo

Urban poverty and water shortage go hand in hand: Makoko, Lagos, Nigeria.
Photo taken in 2002. Stuart Franklin/Magnum Photos

Phnom Penh, Cambodia. Photo taken in 2004. John Vink/Magnum Photos

Manila, the Philippines. Photo taken in 2003. Rolex dela Pena/EPA Photo

Cité Soleil, Port-au-Prince, Haiti. Photo taken in 2006. Redux/laif

Kuala Lumpur, Malaysia. David Nunuk/Keystone

Megacities–
the quiet disaster

We visualize water scarcity as a woman walking miles through the desert with a water jug on her head. This image is deceiving, putting a gloss on the infinitely more serious reality. Supplying small village communities in the desert and barren areas is the least of the problems mankind faces in providing itself with clean water. The most urgent and indeed most dire troubles are in huge urban centers like Mexico City, São Paulo, Kinshasa, Mumbai, Calcutta, Lagos, and Dhaka. Water scarcity affects thousands of cities, from megalopolises boasting populations in the millions to countless smaller and middle-sized cities in Asia, Africa, the Middle East, Latin America, and the Caribbean.

Over one billion people, about a fifth of humanity, have to manage without safe water. Over two billion suffer because of unacceptable sanitary conditions.

Water shortage is not just about not having enough water. Water scarcity is a problem that the world community has created and indeed continues to re-create every day. The lack of access to clean water is a story of poverty, repression, negligence, and irresponsibility. It is a colossal but quiet disaster.

Poverty is migrating into the cities. Driven by the hope for work and prosperity, hundreds of thousands, even millions of people are streaming into big cities every year, especially in developing countries. Mexico City alone has grown by two million persons a year for several years now. Roughly half the world's population already lives in cities. United Nations experts claim this figure will increase to three fourths in ten years. Some 340 million people, more than the entire population of the United States today, will then inhabit 20 to 30 huge urban centers with a population of more than 10 million inhabitants each.

Not all these migrants are poor, but today every third city-dweller in the world already lives in a slum, a shanty town, a decayed building, or an emergency shelter. In the big cities of sub-Saharan Africa, this figure is as high as 72 percent and in cities in southern and central Asia at least 58 percent. Experts estimate that, in thirty years, a third of the world's total population will be living in slums.

Here, where poverty, disease, unemployment, crime, and other social problems are linked and aggravate the situation, the lack of safe water is one of the most pressing problems–and one that's almost impossible to solve. Constructing reliably functioning water-treatment and supply systems, and ensuring efficient and proper wastewater disposal, are complicated and expensive ventures. To cope with the annual migration of hundreds of thousands of people into poor districts, a city would have to expand its water and sanitation system each year by as much as the whole water infrastructure in a city the size of Zurich. This means that some 1,500 kilometers of pipeline would need to be laid every year, and many waterworks, sewage plants, reservoirs, and pumping stations built. At the same time, new sources providing several hundred million cubic meters of water would have to be tapped year after year. These new investments alone would run into billions of dollars.

It would cost additional millions to maintain existing networks and this figure would increase over the years. The replacement value of public water utilities in Switzerland, for instance, a country rich in water and with a total population of less than 7 million, amounts to about 80 billion dollars. Annual maintenance and operational costs come to over 2 billion dollars.

Hardly any megalopolis is in a position to foot the bill for such investments and operating costs by itself through taxes and water fees. Even financially able cities in regions with plenty of water, such as New York, Tokyo, and Paris, where public water utilities have grown slowly over about 150 years and immigration has stayed within bounds, reach their economic limits here (→ p. 236).

Most megalopolises are in poor and underdeveloped countries, many of them in regions already suffering from water scarcity. Their biggest problem is not in developing public water utilities but in acquiring additional water resources. Easily accessible sources have long been tapped, and the cost of exploiting additional water resources is very high. These cities need to invest in expensive drilling to get to deeper groundwater, or water has to be brought in through extended networks of canals or pipelines from ever farther away, often over hundreds of kilometers.

Even within these huge cities, water is very unevenly divided. While the water supply to those living in affluent neighborhoods reaches European and American standards, people living in slums can consider themselves lucky if they manage to get hold of a few liters of clean water every day. Most city authorities and local governments are more interested in making their city attractive for investment, international industry, and service businesses, and for tourists and high-profile sports events than they are in fighting poverty. Many administrations are also simply unable to cope. According to UNESCO's 2003 *Water For People—Water For Life* report, good governance—a competent, transparent and democratic style of government—is one of the most important prerequisites for alleviating water scarcity in slums.

It would be a huge step forward if at least existent water supply systems were well maintained, repaired, and regularly replaced. UNESCO's water report estimates that an average of 40 percent of all drinking water is lost from leaking pipes in many African, Asian, and Latin American cities. Governments have other priorities. They make costly new water resources available to industry and agriculture, leaving urban water facilities to decay. Municipal authorities today are increasing their efforts to privatize public water facilities or transfer their obligations to joint public-private ventures, although not very successfully (→ pp. 430).

For conceivable reasons, there are no official figures for water scarcity in the poor districts of many megalopolises. Municipal authorities often don't even know how many people live in their slums and they prefer to ignore illegal settlements altogether. However, studies carried out by WHO, UNICEF, and aid organizations in some cities enable us to at least estimate the scope of the disaster.

Only half of the six million people living in the poor areas of Karachi in Pakistan have reliable access to safe water. In the slums of Chittagong, Bangladesh, a city of 3.5 million, only one in every four households has running water. Some 200,000 people get their water from one of 600 public hydrants on the street, but one million inhabitants get their supply from dirty ponds, canals, or water barrels used to collect rainwater.

In Dhaka, the capital of Bangladesh, half the people living in the slums have to walk at least half an hour to reach safe water. In Nairobi, Kenya, where over half the population lives on an area covering 6 percent of the city, only 12 percent of slum-dwellers have access to public water supply. Over 80 percent fetch their water in canisters from one of the private water kiosks, paying 4 to 5 Kenyan shillings for 20 liters of water. This is ten times more than the price paid by those living in exclusive residential districts where houses are connected to the municipal water supply. In the capital of Mali, Bamako, private vendors sell water for thirty-five times more than the municipal waterworks charge. A lucrative business—to date the city has not extended water pipelines to the slums.

UNESCO's water report points out that the lack of safe water is equally dramatic in thousands of middle-sized and smaller towns in rural Africa, Asia, and the Middle East. Many people in towns are just as poor as slum-dwellers in megalopolises, and infrastructures here are usually even more rudimentary and public funds even more limited. Many of these towns are a long way away from areas subsidized by the government and carry no weight, either politically or economically. Nor do they have important industrial corporations exerting influence to improve their situation. They are left to fend for themselves and the international community is hardly aware of their plight. These towns' inhabitants are the silent victims of a quiet disaster.

First aid: Connected to a water main, this water kiosk in Makindu in Kenya serves as the source for clean drinking and washing water for 15,000 people within a radius of 10 kilometers. Project of Welthungerhilfe (German agro action). 2005.
Wernet/laif

Daily routine: The arrival of the water tanker dominates the day of many housewives who live in poor countries. A single tanker holding 10,000 liters of water is emptied into jars and bowls within minutes. In Mumbra, a small suburb of Bombay. Savita Kirloskar/AVD/CMC

A dangerous journey: Despite the perils involved, most of the inhabitants of Bel Air (a slum plagued by violence in the Haitian capital Port-au-Prince) have to walk several blocks to buy water at one of the few available taps. Photo taken in 2005. Kent Gilbert/AP Photo

Water in times of war: The water supply in Basra in southern Iraq is frequently interrupted. Photo taken in August 2003. Sergei Grits/AP Photo

Where infrastructures disintegrate, water pipes will sooner or later run dry.
Malabo, Equatorial Guinea, Photo taken in 2002. Christine Nesbitt/AP Photo

Water vendors as a substitute for a mains water supply in Jakarta, Indonesia. Photo taken in 1999.
Curt Carnemark/AP Photo

Pueblos Jovenes, Lima, Peru. Photo taken in 2002. Yoshiko Kusano/Keystone

Rain is a blessing: Freshly collected rainwater is perfectly hygienic. Angkor Meanchey, Phnom Penh, Cambodia.
Photo taken in 2002. John Vink/Magnum Photos

Rocinha favela, Rio de Janeiro. Abbas/Magnum Photos

Connected to water supply: worldwide coverage (2000)

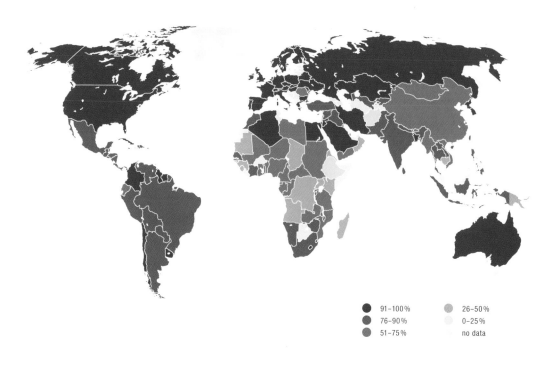

- ● 91–100%
- ● 76–90%
- ● 51–75%
- ● 26–50%
- ● 0–25%
- no data

Water supply in large towns and cities
Average percentage of the population according to type of connection and region

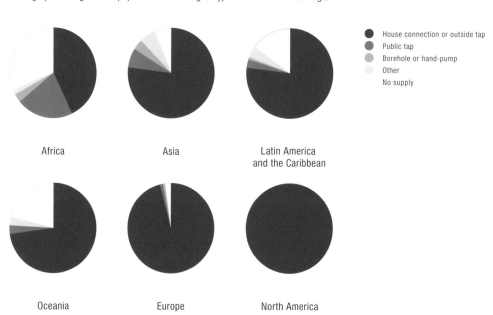

- ● House connection or outside tap
- ● Public tap
- ● Borehole or hand-pump
- ● Other
- No supply

Africa

Asia

Latin America
and the Caribbean

Oceania

Europe

North America

Annual per capita water requirement

For domestic, service company and industrial uses (1990-1995)

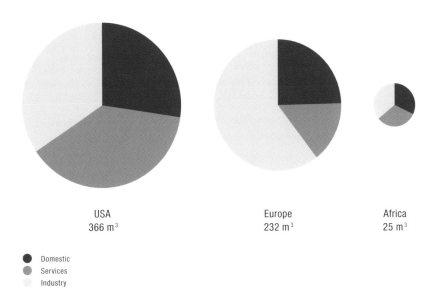

USA	Europe	Africa
366 m^3	232 m^3	25 m^3

- ● Domestic
- ● Services
- ○ Industry

Global Water Supply (1990/2000)

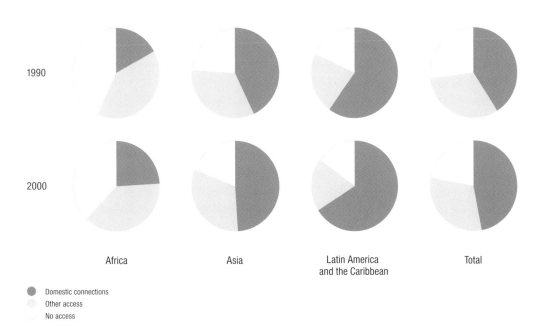

1990

2000

Africa	Asia	Latin America and the Caribbean	Total

- ● Domestic connections
- ○ Other access
- ○ No access

80 percent
in the Third V
to contamir
or a lack of
day, **6,000 |**
children und
of five, **die fr**
related illnes

of all illnesses

/orld are **due**

ated **water**

water. Every

eople, mostly

er the age

om diarrhea-

ses.

Is everything under control? The problem of how to dispose of wastewater is considered solved.
London 2002. Peter Marlow/Magnum Photos

Cleanliness is a duty: highway toilet near Magdeburg. Modrow/laif

A dubious mixture: Even the most advanced sewage treatment cannot satisfactorily deal with the mix of bath and kitchen waste, feces and industrial effluent. Niederrad sewage treatment plant for the city of Frankfurt am Main, Germany.

Filth, excreta, and flying toilets

Two billion people don't have access to clean toilets or washing facilities. In Africa, it is 40 percent of the population, and in Asia as much as 52 percent. The excreta of two billion people pollute backyards and streets every day, seeping into groundwater and contaminating drinking water. Every year over a hundred cubic kilometers of wastewater, filthy with household waste, contaminated by germs, and polluted by detergents and household chemicals, flow untreated into rivers, seas, streams, and ponds. In New Delhi alone, 200,000 cubic meters of household wastewater flow untreated into the Yamuna River every day.

Larger cities and metropolises in developing countries are affected the most as far as wastewater disposal is concerned. Makeshift solutions feasible in rural areas can't be put into practice in cities where many people live very closely together—in small villages and communities people can at least relieve themselves away from wells and water sources.

Reliable figures are scarce. A study by the World Health Organization (WHO) in 116 of the world's cities revealed that not even 20 percent of households in Africa are connected to a sewer system. In Latin America, Asia, and the Caribbean, this figure is no higher than 40 percent. Although 40 percent of slum-dwellers in Dhaka, Bangladesh, have access to toilets and hygiene facilities with sewage drains, 42 percent relieve themselves out in the open. Statistics on Karachi attest that 40 percent of households are connected to the sewer system, but this actually means that 80 percent of households in prosperous districts are connected, while only 12 percent are in poorer districts, where a household can consist of an entire clan. In Nairobi, 94 percent of all people living in slums have no access to proper sanitary facilities. People living in many of these slums dispose of their excreta in "flying toilets," paper or plastic bags that end up somewhere in backyards or on waste dumps.

But these figures reflect only half the reality. Many of the latrines and toilets recorded in statistics are so poorly looked after, filthy, or damaged that they can't be used at all. It is hard to keep a toilet block clean when it is being shared by hundreds, even up to 2,000 men, women, and children, including those who are old and ill. Furthermore, in many poor neighborhoods people have to pay to use publicly or privately operated toilets and sanitary facilities. Although they are included in the statistics, these toilets aren't used because they are simply beyond many people's means. Many poorer families would have to spend up to 15 percent of their total income for every member of the family to be able to use a toilet and wash themselves once a day.

Poor freshwater supply and the lack of safe disposal of excreta and household wastewater are–together with malnutrition–the main causes of epidemics in the Third World. WHO calculates that at any given time, roughly half the population in developing countries suffers from one or more of the six most common illnesses caused by unsafe water and poor hygiene. Some 1.7 million people, including more than 600,000 in Africa, and nearly 700,000 in Asia, die of water-related diseases every year, 90 percent of them children. WHO calculates that up to two million people die annually from diarrhea alone–1.4 million of them children under the age of five.

Child mortality in the Third World is on average ten to twenty times higher than in western industrialized countries. But here too, statistics only barely reflect the full reality. Child mortality in Karachi, for instance, varies from 3 to 21 percent depending on what district the children live in. In Germany, the average figure comes to just 0.5 percent.

According to WHO, about 80 percent of all illnesses in developing countries are water-related. There is a long list of illnesses resulting directly or indirectly from water scarcity, contaminated water, or inadequate sanitary facilities. It includes cholera, typhoid fever, polio, numerous skin and eye diseases, hepatitis A (to some extent), threadworms, and parasites. Nowhere do mosquitoes or gnats, the carriers of malaria and dengue fever, reproduce more quickly than in polluted waters, ponds, sewers, and garbage dumps containing excreta and household waste.

Destitute people in many urban areas in developing countries have no choice but to use water which has already been used for industrial or agricultural purposes. This water almost always contains high concentrations of fertilizers and pesticides, solvents, oils, or heavy metals from industrial production. The pesticide DDT, proven to be seriously harmful to human health and long banned in the United States and the European Union, is still used to combat mosquitoes in numerous countries in Africa and Asia.

Not all water contamination stems from human activity. In Bangladesh, for instance, arsenic has been released from rock in the Himalayas for thousands of years, seeping into deep groundwater in the estuary areas of the Ganges, Yamuna, and Brahmaputra River. This has become the cause of one of the world's biggest health disasters. As long as people consumed drinking water mostly from rivers and shallow wells, their level of exposure to arsenic was not hazardous. But as surface waters became increasingly polluted, international aid organizations in the 1970s and 1980s helped install more than a million new hand pumps to bring up ground-water as an alternative.

People stopped using river water and the number of cases of diarrhea fell drastically. But at the same time, some 30 to 50 million people were unknowingly consuming groundwater that was heavily contaminated with arsenic. Not until the 1990s, two decades later, when people first began to die from the long-term consequences of slow poisoning, did the public become aware of the problem. Today one million people are afflicted with chronic arsenicosis. WHO estimates that up to 300,000 Bangladeshis will die of this illness in coming years. Millions of people in China, Vietnam, India, Pakistan, Ghana, Brazil, Argentina, Chile, Britain, and the United States also drink water containing arsenic.

In comparison to these countries with their enormous problems, the wealthy industrialized nations of Europe and North America are in a very comfortable position. As far back as Byzantine and Roman times, larger cities like Constantinople (present-day Istanbul) and Rome built aqueducts and canals to supply their populations with fresh drinking water. In the mid-19th century, scientists discovered that typhoid and cholera became epidemic diseases when people used water contaminated with excreta. Larger cities then began separating the provision of drinking water from the disposal of wastewater.

Chicago, for instance, constructed longer and longer pipelines into Lake Michigan so that the city could still get clean drinking water from close to shore. When this plan didn't work, the city in about 1900 built the Illinois Waterway. This was the biggest water-engineering project carried out before the construction of the Panama Canal. It reversed the directions of the Chicago River and the Calumet River so that wastewater from the city and its huge slaughterhouses no longer flowed into Lake Michigan but was conducted through canals and pipelines into the Illinois River, and from there into the Mississippi—much to the displeasure of down-stream states. They took the city of Chicago to court, forcing it to cleanse its wastewater.

The most important change in the history of hygiene was the invention of the water closet. As long ago as 1596, a Sir John Harrington invented a kind of flushing system. The water closet itself was patented by its inventor, Alexander Cummings, in 1775. But even in Europe and North America, the flushing toilet did not become a regular feature in homes until the end of the 19th century.

The first sewer systems were built in London, Paris, and many other European and American cities. They collected wastewater and excreta from households and discharged them untreated into rivers and lakes. This was successful only to a limited degree. In 1858, British members of Parliament still had to protect themselves from the "great stink" rising from the Thames by soaking cloth in calcium hypochlorite and hanging it from windows in the Houses of Parliament.

By the early 1920s, most cities and larger towns in Europe and North America were at least supplied with drinking water and had sewer systems that worked relatively well. This later became the case in European districts of Asian colonies as well. At the same time, many cities chose to clean their wastewater by using it to irrigate fields in an attempt to keep the waste away from rivers. Indeed, Berlin in 1927 was still disposing of all its wastewater, some 182 million cubic meters a year, in this way. The city's Spree River was already intolerably polluted and the few treatment plants in existence had been shut down due to capacity overload.

The era of modern sewage-treatment plants began when British engineers developed a process for purifying water by using activated sludge bacteria to consume and remove organic matter. The first sewage treatment plants in Germany were already being built in the 1920s. Washington, D.C.'s first facility went into operation in 1934, and Moscow's at the end of the 1930s. Many further improvements and inventions were still needed before treatment plants were able to effectively eliminate phosphorus and nitrogen from wastewater. Comprehensive wastewater disposal and purification has been in place in developed countries only since the 1970s. Milan and Brussels, for example, reached this standard only after the turn of the millennium.

The first sewage-treatment plants were designed solely with the intention of preventing acute health problems and improving aesthetics. In the last few decades, public attention has increasingly focused on ecological aspects such as keeping lakes and oceans clean, saving fish stocks, and protecting biodiversity and ecosystems. Interestingly, industrial businesses have also pressed for sewage plants to be built because they depend on clean water for many of their production processes. At the same time, they have been less willing to clean their own wastewater.

Even modern water supply systems are beginning to reach their limits. Careless water consumption and the discovery of new, previously unknown contaminants in drinking water are forcing even those cities with well-developed water supply and wastewater disposal systems to continually make improvements. They often have to rethink their disposal management policies, taking the latest scientific findings into account.

Some experts suggest, for instance, that urine be separated from common domestic wastewater in "no-mix" toilets and disposed of separately (→ pp. 380). Although urine accounts for only 1 percent of wastewater, its ingredients contaminate all of it. Almost 90 percent of the nitrogen and 60 percent of the phosphorus in wastewater come from urine. The no-mix process could drastically reduce these figures and even reclaim most of the phosphorous. A useful side effect—recent estimates indicate that the world's workable reserves of phosphorus, one of the most important fertilizers used in agriculture, will run out in about a hundred years.

Furthermore, pharmaceuticals, hormones, and other bioactive substances are increasingly entering wastewater through urine. Traditional treatment plants can only partially remove them. We don't have nearly enough information yet on the effects of these hazardous substances on animals and humans. Concerned scientists warn that these new pollutants could pose one of the greatest challenges to keeping water clean over the next few decades.

Experts around the world agree that a new concept of waste disposal is needed to solve these problems. Wherever possible, polluting substances must be replaced by harmless ones, or at least removed from water as closely as possible to the actual point of contamination. Once they enter rivers, lakes, and groundwater, it is nearly impossible or extremely expensive to remove them.

THE FOUR CORNERSTONES OF MODERN URBAN WATER MANAGEMENT:

1. Protecting resources
Water should be kept free of chemicals and pathogens. Sewage must be treated as closely as possible to the point of contamination. In addition, regulations for agriculture and industry must be stricter.

2. Purifying drinking water
Technologies must be adapted to specific local conditions and pollutants, if need be according to the multi-barrier principle.

3. Maintaining infrastructure
Good network management, regular monitoring of pipelines, long-term maintenance, and renewal, the use of more durable materials, and the stabilization of water pressure should ensure that an existing system of pipes remains intact and functions well.

4. Replacing household installations
Regulations must ensure that sanitary facilities in individual households meet current best standards. This includes installations and the replacement of old pipes made of hazardous materials. In areas where water is scarce, supplying drinking water and lower-quality gray water through separate pipes can greatly contribute to drinking water being used more economically and efficiently.

Skepticism is understandable: Where sanitary facilities are few and far between, the inhabitants must dispose of their waste water at the nearest open space. Dakar, Senegal, 2005. Nic Bothma/EPA

Water stress: Water supply and wastewater must be kept separate under all circumstances.
Failure to do so always has grave consequences for the population. In Dhaka, Bangladesh. Hoogte/laif

Overcrowding, poverty and neglect on the part of the state: Whenever services break down, wastewater becomes a hazard. Controlling stickworms in a sewer in Lagos, 2002. Stuart Franklin/Magnum Photos

Hygiene in housing estates in large towns and cities

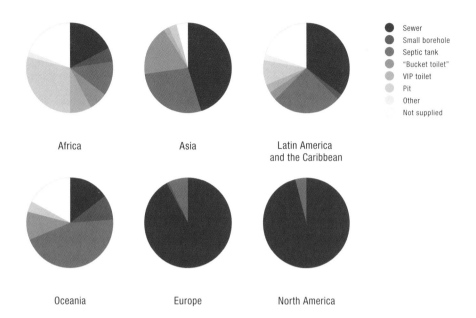

- ● Sewer
- ● Small borehole
- ● Septic tank
- ● "Bucket toilet"
- ● VIP toilet
- ● Pit
- ● Other
-   Not supplied

Africa

Asia

Latin America
and the Caribbean

Oceania

Europe

North America

Sewage systems
Connections (as a percentage)

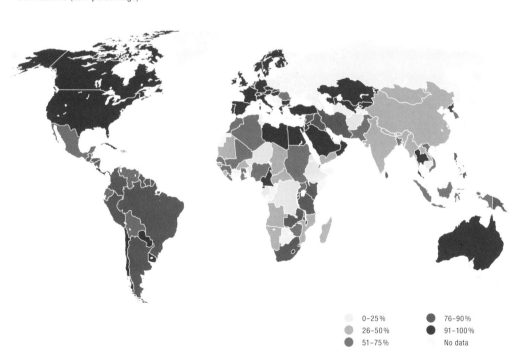

- ○ 0–25%
- ○ 26–50%
- ● 51–75%
- ● 76–90%
- ● 91–100%
-   No data

Sewage disposal
Degree covered by diverse systems

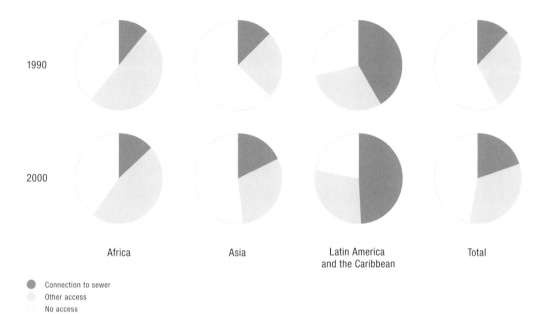

1990

2000

Africa   Asia   Latin America   Total
                              and the Caribbean

- Connection to sewer
- Other access
- No access

Households' water consumption
Typical pattern in an industrialized country in 2003. Large quantities of water are contaminated by faeces in the toilet.

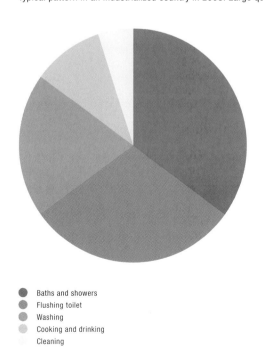

- Baths and showers
- Flushing toilet
- Washing
- Cooking and drinking
- Cleaning

In the USA, c
consumes a
300 liters a
personal us
for **Europe i**
and for Afri

each person pproximately **day** for . The figure **s 160 liters, ca 30 liters.**

Islands are laboratories of resource management: The natural limits of water consumption are more apparent here as people have to make do with the local water supplies. Key West, Florida. Contantine Manos/Magnum Photos

Many towns and cities are in no position to adequately repair their deteriorating water mains systems.
A shower alongside a burst water pipe in Calcutta. Photo taken in 2005. Ilse Frech/Lookat

A booster pump in the home provides the water pressure needed for a good shower. In London, the private water company Thames Water has considerably lowered the water pressure in order to spare the fragile pipe network, which has been neglected for years. Photo taken in 2003. Peter Marlow/Magnum Photos

An abundance of water? Hotel pool on Mauritius. Abbas/Magnum Photos

Closed due to water shortage? Swimming pool in Havana, Cuba. Alex Webb/Magnum Photos

A stark contrast: These housewives on the outskirts of the Indian computer boomtown Bangalore have to carry home the water for their families from public taps. Photo taken in 2006. Manjunath Kiran/Keysteone/EPA

How much water does a person need?

In the *Neue Zürcher Zeitung*, a reporter tells the story of Iris Libokoyi, a young Nigerian woman in Nairobi. Four times a day, Iris heads for a water kiosk with one of her four plastic canisters. Each trip takes 45 minutes. Each day she carries back 80 liters of water so that her family of three has enough water to survive. Iris Libokoyi lives in Kawangware, one of the better slums of Kenya's capital. Well over a billion people in the world live just like her.

Just a few kilometers away from Kawangware, in the exclusive residential district of Lavington, unlimited amounts of running water are available 24 hours a day, and except when heavy rainfall is too much for the watersupply system to manage, it is sufficiently clean. Each person living here uses two or three times as much water as Iris Libokoyi's entire three-person family. Lavington enjoys the same water-supply standards as in Europe or North America.

The average per capita consumption of clean water in Germany is about 130 liters a day. In the United States, it is between 200 to 300 liters, depending on the origin of statistics. Germany's Federal Environmental Agency has calculated that the average consumer in Germany needs only about 4 liters a day for drinking and cooking purposes, but uses almost 50 liters for personal hygiene, washing hands, and taking showers or baths, 40 liters for flushing toilets, 18 liters for washing clothes, 8 liters for washing dishes, and another 10 liters for cleaning, watering gardens, and washing cars.

If European standards for clean-water consumption were applied to the same extent worldwide—not an extravagant idea at first glance, since this standard is taken for granted by most people in more prosperous countries—many of the most highly populated countries would have to triple their drinking and household-water resources today, if not increase them four- to fivefold. In the three decades to come, these resources would have to increase yet again because by that time populations in these countries are expected to have doubled.

Sewers would of course also have to be developed to the same standard, because drinking-water supply and sewage disposal go hand in hand. Dirty water has to be collected and purified, otherwise there can be no safe drinking water.

"The shortage of fresh and clean drinking water," a United Nations human rights commission report says, "is the most serious threat that the human species has ever encountered." Since the 1970s, a series of international conferences have dealt with the water crisis and its social, economic, and financial consequences, as well as repercussions on health (→ pp. 504). At the United Nations Water Conference in Mar del Plata in 1977, governments agreed on a plan of action that aimed to provide everyone with access to drinking water and sanitary facilities by 1990. This was an unrealistic target. Some 650,000 people every day would have had to be connected to public water utilities during the entire Water for Life Decade. This figure adds up every month to more than the entire population of New York City.

The governments represented at the International Conference on Water in Delhi in 1990, and the UN Children's Summit in New York City the same year agreed to set the same ambitious goal, but pushed the target forward to 2000. By 2000, conditions had indeed improved in many places. Today every sixth person in the world still has to manage without clean drinking water and every third person has no access to sanitary facilities. One of the millennium development goals agreed upon at a UN summit in September 2000 sounds much more modest. The number of people "without sustainable access to safe drinking water" is to be halved by 2015. Furthermore, there should be considerable improvement in living conditions for at least 100 million slum residents by 2020. But experts believe even these goals can't be met without radical measures being taken.

A number of conferences hosted by WHO, UNICEF, the UN Development Program (UNDP), and the UN Human Settlements Program (HABITAT) went beyond the mere declaration of general goals. Taking health and sociopolitical aspects into consideration, water experts attempted to determine what the minimum "sufficient" supply of safe drinking water and the minimum for "appropriate" sanitary facilities should be. Developing countries lack the financial means, infrastructure, and necessary water resources to establish a water supply system with European or North American standards.

The Global Water Supply and Sanitation 2000 *Assessment Report* proposed that a reasonable minimum daily quota was 20 liters per person. The nearest available water and sanitary facilities should not be farther than a kilometer away from any household. Of course, defining such minimum standards does not cover everything since access to safe water and toilets also depends on a number of other factors like whether the supply of water is always reliable all year long, whether water is clean enough, and whether it is affordable for the people concerned.

The UN World Water Development Report states that sanitary facilities must also fulfill a number of requirements for them to be practicable in the long term and acceptable to those using them. Cleanliness, regular maintenance, and reliable sewage disposal are determining factors in slums where tens or hundreds of thousands of people live in overcrowded conditions. The number of people using toilets and washing facilities at the same time should not be underestimated. Also, it is crucial for women and children to be able to safely reach and use these facilities at all times. The privacy needed for personal hygiene in different cultures should also be ensured.

The focus of discussion has shifted considerably since the early 1990s. International conferences such as the Global Water Partnership or the World Water Council, dominated by individual governments, the private sector, and international financial institutions like the World Bank and the International Monetary Fund, are concerned primarily with defining water as a commodity, believing the water crisis should generally be solved through privatization (→ pp. 430).

A broad front of opposition including NGOs, unions, and the governments of developing countries has formed, opposing this strategy for commercializing the water sector. These groups particularly criticize the fact that private enterprises aren't able or willing to make investments unless they profit in the long run. They fear that privatizing water services will be of little benefit to those most in need.

In most cases, the approach taken by development aid organizations is completely different. They promote self-help solutions encompassing the participation of those directly concerned. Since most developing countries lack funds for building comprehensive water distribution systems, solutions should be tailored to specific local conditions and take social and cultural factors, as well as financial ability into account. The most successful projects to date have been those initiated by community groups or local people, developed and carried out in cooperation with external experts, and then maintained by the users themselves. There are various options for working with private companies. These are currently being discussed under the heading of public-private partnerships (→ pp. 458).

In Pune, India, for instance, where 40 percent of the 2.8 million inhabitants live in any one of altogether 500 slums, three private aid organizations, in cooperation with municipal authorities, developed a plan to build 114 sanitary facilities with 2,000 toilets for adults and 500 for children. Instead of setting up the usual makeshift toilets, technicians built solid, well-lit, well-ventilated, and easy-to-clean cubicles. Large reserve tanks and septic tanks now provide enough water all the time, guaranteeing that the facilities remain functional even if disposal problems occasionally arise. To ensure safety and a minimum amount of privacy, separate areas have been created for men, women, and older children. Users have participated in construction and maintenance, and housing has been made available to maintenance and cleaning personnel.

Even makeshift solutions can at times greatly improve the quality of drinking water. Swiss experts in Bolivia successfully tried out a simple procedure to disinfect dirty water by using a method developed by researchers at the Swiss Federal Institute of Aquatic Science and Technology (Eawag). Water contaminated by bacteria is put into transparent PET plastic bottles and exposed to full sunlight for several hours. The combination of heat and sunlight destroys 99.99 percent of viruses and bacteria (→ pp. 378).

The success of such initiatives shows that ideas imposed from above, in which governments, international financial organizations, and large commercial businesses proposing costly and technically complicated large-scale projects that tend to exceed budgets, and forcing them upon populations with no say, are less likely to achieve globally stated goals. Small local projects lead to better and more efficient, if imperfect solutions, at least in the short and medium term. They can gradually be expanded later on. Their greatest merit is to restore people's dignity and sense of responsibility by allowing them to be actively involved in their own affairs, taking their future into their own hands.

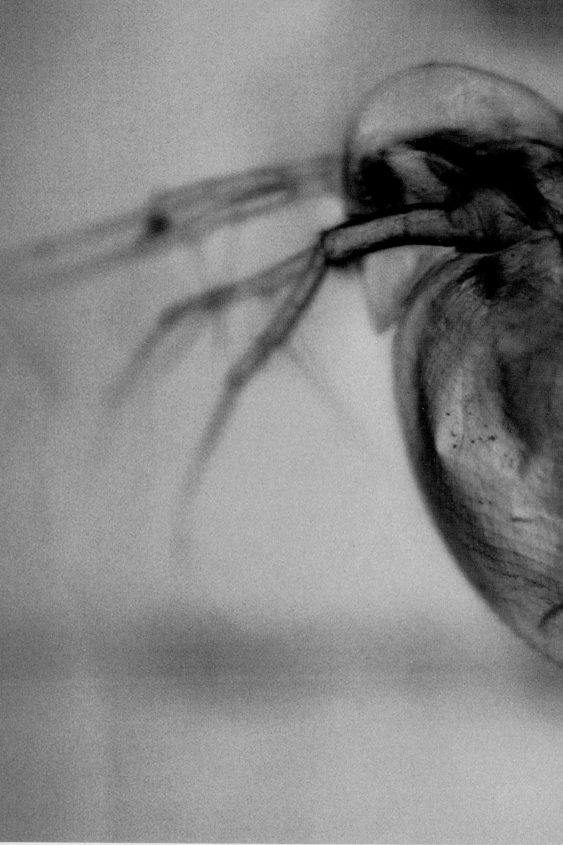

Taster: Zurich's public water company uses the water flea Daphnia hyalina, which reacts extremely sensitively to noxious substances, to check the water quality. Christian Rellstab/Eawag

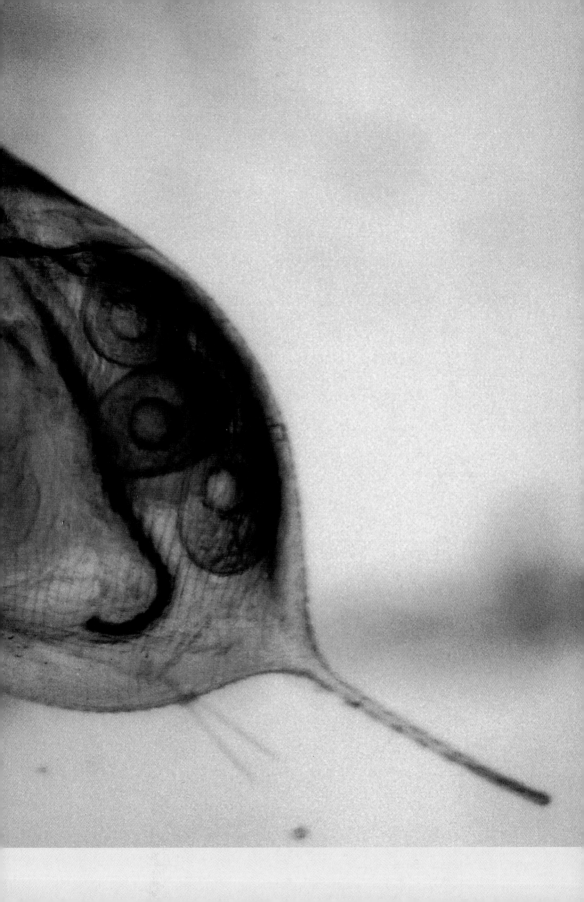

The quality of life: Anyone who can drink water straight from the tap without any qualms still belongs—in global terms—to a minority. Holzenleuchter/Stern/laif

Purity and reliability: Vienna's waterworks are upheld as the model of an exemplary water supply, safeguarded by excellent spring water from remote mountains and a constitutional law mandating that the waterworks remain under full municipal control. Water reservoir in Vienna. Fritz Schmalzbauer/Wiener Wasserwerke

Poisoned waters

Up to 500 million tons of industrial wastewater and sludge seep into groundwater or flow into rivers, lakes, and the world's oceans every year, polluting them with heavy metals, solvents, detergents, oils, fats, acids, bases, radioactive substances, and other chemicals. Added to these are more than one hundred million tons of fertilizers and several million tons of pesticides used in agriculture, and all kinds of chemicals present in another 700 or more million tons of domestic wastewater discharged into natural waters largely untreated, at least in developing countries. This chemical pollution of the environment has become one of the most urgent global problems.

The consequences are devastating. The Chinese environment minister has admitted that more than 80 percent of rivers in China are so contaminated that they are no longer suitable for drinking water or washing. The situation is similarly alarming in many other emerging countries. Even in highly developed nations in the northern hemisphere, the chemical pollution of rivers and lakes is still an acute problem. In the United States, two out of every five rivers, including nearly all of the larger ones, are so contaminated that health authorities warn inhabitants not to swim or fish in them. The one-time hope that the water cycle would act as a global purification plant and the oceans would serve as universal garbage dumps for modern civilization has long since proven to be a fallacy. Many countries are still not prepared, for economic reasons, to draw the obvious conclusions.

Water is not only directly contaminated by polluted waste-water–industrial end products themselves also endanger nature and water. Most daily items of practical use, from household appliances and textiles to detergents, construction materials, paints, automobiles, computer chips, and toys, also contain synthetic chemicals such as flame retardants, preservatives, solvents, softeners, pesticides to combat fungi, bacteria and algae, and other problematic substances. Many consumer items release harmful chemicals while they are being used, many of these in turn reaching water. The disposal of these products at waste-disposal sites and landfills is an additional threat to the environment since toxic substances almost inevitably come into contact with seeping rainwater and ultimately contaminate the groundwater below.

An effective and sustainable strategy against this almost imperceptible and hugely underestimated contamination of the environment is currently not in sight. There are however some quite promising approaches today for reducing the enormous amounts of industrial contaminants released to the environment in wastewater. Questionable materials can be replaced by harmless substances and industrial wastewater can be recycled. Several countries have banned the use of some of the most environmentally dangerous chemicals, and a few substances are even banned all over the world. But a global policy on chemicals which adequately balances environment and human health concerns against economic and political interests is barely discernable. Such a policy poses a highly controversial issue in poor developing countries, ambitious emerging countries, and powerful industrial nations alike, since it would directly conflict with today's dominating economic and industrial interests.

A plume of wastewater coming from a paper mill near Chongqing, a city with 8 million inhabitants in China.
Less visibly, industrial wastewater also heavily pollutes rivers with chemicals in Europe and North America.

Bruno Barbey/Magnum Photos

Environmental problems caused by mining: More rubble, debris and sediment are washed into rivers each year due to the extraction and processing of metals and precious stones than due to natural erosion. Furthermore, noxious substances, some of them highly toxic like cyanide, are released into the environment. Ruby and jade mining in Myanmar. Steve McCurry/Magnum Photos

Hundreds of thousands of tons of insecticides, herbicides, fungicides and other biocides are applied directly into the environment, contaminating lakes and rivers worldwide. Insecticide-spraying in the Rio Grande Valley, Texas. Garry D. McMichael/Keystone/Photo Researchers

The aim is to purify industrial effluent so that it can be reused in production. To this end, toxic and persistent substances must be replaced by less hazardous products. A detergent factory by the River Thames, Essex, Great Britain. Keystone

As many as
tons of **indu**
are discharg
lakes, and s
40 percent
are **unsuital**
for fishing o
drinking wat
figure is **ove**

500 million
strial waste
ed into rivers,
eas. In the USA,
of all rivers
ble for bathing,
as a source of
er. In **China**, the
r 80 percent.

Useful products also have their drawbacks: Pharmaceuticals potentially pollute the environment when they are excreted by patients. Some of these substances can be found in rivers, groundwater and even in drinking water. Barleben, Sachsen-Anhalt. Photo taken in 2006. Martin Ruetschi/Keystone

The more resources are consumed, the more lakes, rivers and groundwaters suffer from water abstraction and pollution. There are alternatives, however: Newsprint can be produced from recycled paper in a closed water cycle. Sigi Tischler/Keystone

Consumer goods which are available in unlimited quantity, cheaply, and at all times, require industrial mass-production systems–often resulting in heavy pollution of the aquatic environment. This factory in Shanghai produces 80,000 meters of cotton yarn a day. Gao Feng/EPA Photo

Insidious poisoning

No production facility can manage without water, whether it is used directly as a product ingredient, for cleaning and rinsing, or for cooling during and between production processes. Industry uses approximately 20 percent of the total volume of water consumed annually, twice the amount used in the households of the six billion people on our planet.

In highly developed nations, the industrial use of water amounts to nearly 60 percent of total water usage, compared to only about 10 percent in developing countries. This difference is an indicator of the explosive growth expected for the future. Thanks to more efficient and water-saving production technology, the industrial consumption of water in highly developed nations has increased very little, and indeed in some countries it has even dropped. Some of the largest and most highly populated emerging countries such as China and India are striving for industrialization by every means. They are developing huge industrial complexes that produce for overseas markets, and, most importantly, their own as yet largely untapped domestic markets. Growing prosperity and the soaring demand for consumer goods have triggered a boom which will cause the industrial use of water to dramatically increase in the next decades. The United Nations Industrial Development Organization (UNIDO) estimates that the industrial use of water around the world will double by 2025.

The growing demand for water is a problem, but what really worries experts is that most production processes result in water pollution. Be it mining or metal processing, the dying of textiles, the tanning of leather, the bleaching of paper, the production of chemicals, plastics, or pharmaceuticals, the processing of foods, or the manufacturing of construction materials–wherever goods are industrially processed, huge amounts of water are used and often contaminated by chemicals. Industrial wastewater containing all kinds of chemical waste and residue is discharged into rivers. In many countries this wastewater has been insufficiently treated, if at all. This is particularly the case in large industrial centers where hundreds or thousands of large and small factories manufacture and process a wide variety of products. The problem is exacerbated when such industrial areas are located in urban centers. Here the receiving waters–rivers, lakes, and seas–are also burdened with millions of peoples' largely untreated sewage.

Some 400 million tons of synthetic chemicals alone are manufactured each year, an amount that would fill a convoy of trucks ten times as long as the distance from Paris to Beijing. This is a staggering figure, even if only a small percentage of these chemicals poses an environmental hazard and reaches aquatic systems.

Many of the by-products of manufacturing processes, heavy metals, solvents, and detergents—found in effluents and other industrial waste—end up in water. Conservative estimates place figures at 300 to 500 million tons of waste material discharged every year into rivers, lakes and oceans all over the world. This amount will quadruple by 2025 unless there is a revolution in production technology followed by the consistent enforcement of stringent environmental laws.

The consequences are devastating. In the United States, almost 40 percent of rivers are so polluted that health authorities advise against swimming or fishing in them, not to mention drinking their waters. Every day, 40 million tons of untreated household and industrial wastewater flow into the Yangtze, China's largest river. A fourth of all rivers in Poland are so polluted that their waters can no longer be used, even for industrial purposes. A study by the Canadian ministry of the environment revealed that 85 percent of the water samples from tributaries of the Saint Lawrence River, officially regarded as clean, still contain significant levels of ammonia, phosphorus, aluminum, arsenic, barium, mercury, PCBs, dioxins, furanes, cleaning chemicals, polyaromatic hydrocarbons, and other organic and inorganic pollutants.

Every year, 130 million tons of fertilizers (nitrate, phosphate, and potash) seep into groundwater or flow into rivers, lakes, and oceans, not to mention the pesticides, herbicides, insecticides, and fungicides that join them there. Some of these are so toxic, like the insecticide DDT, that they have been banned in most western countries for a long time. They are still widely used in many developing countries. Paraquat, a highly toxic herbicide in use since 1961, causes serious damage to the liver, kidneys, heart, and lungs. In spite of strict environmental legislation, its use is still tolerated in the European Union even though it is toxic.

Although water is polluted from local and regional sources, its contamination poses a universal challenge. Today pollution is evident everywhere: in highly developed industrial nations, in the emerging nations of Asia and Latin America, and in the poor developing countries of Africa and Asia.

In terms of quantity, industrial wastewater lags far behind household wastewater and agricultural runoff, but the hazard it presents is a question of toxicity rather than volume. Some chemicals and substances are so toxic that even small amounts can cause extensive damage. Others are extremely dangerous because they degrade very slowly or not at all.

Heavy metals are among the most hazardous environmental pollutants. They are used and released in many production and extraction processes such as mining and the metal, petroleum, paper, and plastic industries. They are non-degradable and spread by wind and water, contaminating the environment, the food chain and finally humans.

One of the first environmental disasters caused by heavy metal ions was Minamata Disease, which in 1953 struck at least 12,000 inhabitants of a coastal town in Japan, afflicting them incurably with paralysis and serious vision and hearing impairments. Between 1955 and 1959, almost every third infant born in Minamata had mental or physical disabilities. Some 3,000 people have since died from the long-term effects of poisoning. Mercury compounds in wastewater released from the Chisso fertilizer and plastic-production facility were behind the disease, which severely affected the central nervous system. The chemicals discharged into the environment were well-known, but the chemical reactions induced by natural microorganisms, leading to the production of highly toxic methyl mercury, were unknown. Methyl mercury accumulated in shellfish, a staple in the local population's diet.

The Japanese ministry of health uncovered the source and cause of the disease in 1959, but Chisso continued dumping wastewater containing mercury until 1968. In 1973, courts ordered the chemical company to pay compensation to its victims. The highest court of appeals acquitted the company in 1992 because it reasoned that the toxicity of the compounds was unknown at the time the ailment broke out. This doubtful argument demonstrates the essence of the problem in enforcing legislation on chemicals—as long as a certain substance has not been proven to be toxic, it is considered harmless. Research on the toxicity of substances is rarely carried out before some kind of damage has been done.

Although the toxicity of many heavy metals is now well-known, countries bordering the North Sea, most of them with relatively strict environmental legislation, still deposit 4,500 tons of lead, 2,000 tons of copper, over 1,000 tons of chromium, more than 100 tons of cadmium, and 64 tons of mercury through rivers and the atmosphere into the North Sea every year. No one knows if these pollutants can ever be removed from the coastal marine environment.

Some of the most hazardous industrial toxic substances are persistent organic pollutants (POPs). These substances take a very long time to degrade, spreading throughout the world in the atmosphere and ocean currents. In contrast to more readily degradable substances, POPs accumulate in plants, animals, and humans, either directly or through the food chain. Often years or even decades go by before their effects become clearly visible as diseases or ailments that can be incurable.

Even less research has been done on bioactive chemicals such as antibiotics, hormones, vaccines, pain killers, psychotropics, and the degradation products of these substances, which pollute water through household and animal-farm runoff. They can have various effects on organisms and some of them are thought to cause genetic mutation. Currently no one can guess how these relatively new substances will affect aquatic life and humans over the long term. The presence of these chemicals in bodies of water is increasing heavily.

Depending on the method of counting, the industrial sector currently produces between 50,000 to 100,000 different chemicals, many of them coming into contact with air or water at some point. The list grows longer by another 1,500 new chemicals every year. Most of these are probably harmless, but right now we don't know which. We know even less about how the mixture of all these chemicals will affect the environ-ment and living organisms.

Environmental scientists and some politicians have recognized that chemicals must be controlled at the source of contamination. Sound environmental measures and stringent legislation must prevent dangerous chemicals from ever reaching bodies of water. Once they are in the water, the job of removing them borders on the impossible.

From its very beginnings, the manufacturing industry has settled along water. Originally it was drawn to rivers by their hydropower potential. Nowadays, the inexpensive disposal of manufacturing wastes via wastewater is the most important factor. BASF chemical factory, Ludwigshafen, Germany. Mathias Ernert/Keystone

Industrial regions

● Main industrial regions

Industrial consumption of water
Quantity of water consumed by industry in 2000

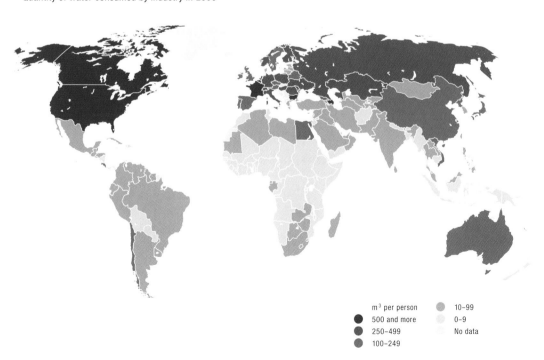

m³ per person
● 500 and more
● 250–499
● 100–249
● 10–99
● 0–9
　 No data

In 2005, mor
people had
tap water fo
in China, wh
an explosion
factory, high
chemicals f
Songhua Riv
source of dr

e than **6 million**

o go without

one week

en, following

at a chemicals

ly toxic

owed into the

er–the **only**

inking water.

The environment and human rights are often sacrificed to the dictates of global competition. Refuse is simply flushed out the backdoor without any regard for the health of the population or any scruples about contaminating drinking water resources. Wastewater from a paper factory in Dongxiang in the province of Jiangxi in eastern China. AP Photo

Atolls of steel: The French oil and gas platform Frigg is one of a group of six platforms that are supposed to pump at least 230 million cubic meters of natural gas out of the seabed to satisfy Western Europe's thirst for energy. Under such conditions, it is almost impossible to avoid polluting the sea. Jean Gaumy/Magnum Photos

More than just collateral damage: It will take quite a few decades for the fauna off the coast of Alaska to recover from the damage caused by Exxon Valdez in 1989. Paul Fusco/Magnum Photos

Ten years after Exxon Valdez, the oil tanker Erika broke up, bringing fishing along the Spanish and French coasts to a standstill. Jean Gaumy/Magnum Photos

Oil pollution caused by Exxon Valdez in the Prince William Sound, Alaska in 1989. Paul Fusco/Magnum Photos

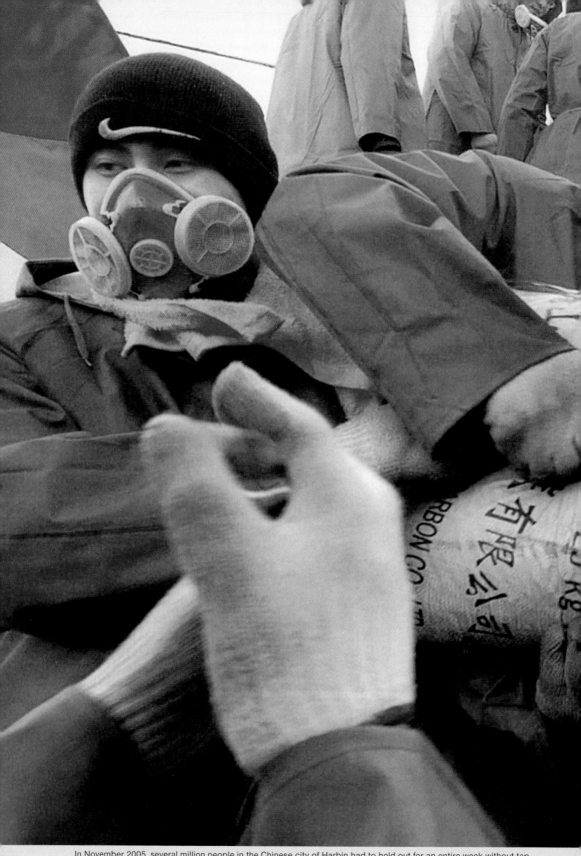

In November 2005, several million people in the Chinese city of Harbin had to hold out for an entire week without tap water because the river supplying their drinking water had been contaminated by an accident involving highly toxic chemicals. A soldier stocks up active carbon intended for the removal of toxins from drinking water. AP Photo

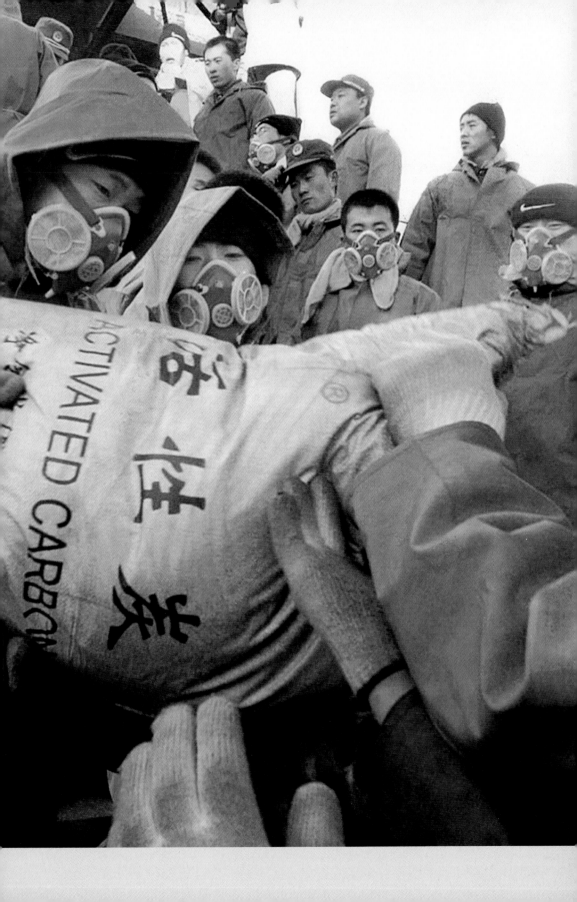

Russian roulette

On the night of November 1, 1986 a fire broke out in a storage facility at the Sandoz chemicals plant in Schweizerhalle on the outskirts of Basel. In the runoff from the water firefighters used to extinguish the blaze, approximately 30 tons of agricultural chemicals were washed into the Rhine, among them insecticides and fungicides. The spill devastated the river's ecosystem for 400 kilometers downstream, turning the Rhine into a biologically dead body of water.

Even before this disaster happened, the Rhine was so polluted that it had been thought of for decades as Europe's sewer. The third longest river in Europe, the Rhine flows through one of the most important industrial regions in the world. Several of the world's largest chemical plants have been active there for a long time. The Ruhr valley region has been the center of German steel and iron production since industrialization began. Potassium mines in France caused the salinity of the Rhine to increase sixfold between 1880 and 1960. Numerous manufacturers using water for their production processes built facilities along the Rhine, and other industries profited from cheap river transportation. Until the 1970s, the sewage from 50 million people flowed mostly untreated into the Rhine and its many tributaries. At the same time, the Rhine was the drinking water reservoir for several large cities and hundreds of smaller towns.

A halt was called to fishing in 1958, some twenty-eight years before the Sandoz disaster. For all practical purposes, the river was dead by 1970. Although Switzerland, Germany, France, Luxembourg, and the Netherlands in 1950 had already founded an International Commission for the Protection of the Rhine (ICPR) against pollution, it was not until thirteen years later that a first agreement, the Treaty of Bern, was signed. By 1976, other agreements had been drawn up, including a treaty to protect the Rhine against chemical pollution. First steps were taken in the 1970s when numerous sewage-treatment plants were built along the river, but it wasn't until the shock of the Sandoz disaster that ministers of the ICPR states agreed to enforce a joint plan for action.

Geopolitics for decades delayed the drawing up of a similar international treaty to protect the Danube, Europe's second-longest river. An agreement was finally signed in 1994, but did not go into effect until 1998, some forty years after the Rhine treaty. No agreement could be made earlier because some of the thirteen countries along the river, notably eastern-bloc states, showed no desire before 1989 to clean up their industrial wastewater. Some of these former communist republics still require foreign aid to build the sewage-treatment plants they need. In the Danube's upper reaches in Germany and Austria, the quality of water has greatly improved since the mid-1990s. But fertilizers and pesticides are still being detected at concentrations above legal limits. The Danube remains severely polluted in Slovakia, Hungary, and Romania. Its waters are still a living lexicon of every kind of hazardous substance.

The international early-warning system along the Danube only recently proved to be ineffective, and emergency plans non-existent. When a waste dam at the Aurul gold mining facility near Baia Mare ruptured in January 2000, spilling more than 110,000 tons of highly toxic cyanide-contaminated water into the Tisza River and the Danube, it took the Romanian government weeks to abandon its strategy of downplaying the incident and to admit the true extent of the disaster. Even ten days after the event, the owners of the mine, the Australian mining company Esmeralda Ltd., were still denying that their Romanian subsidiary was in any way responsible for contaminating these rivers.

Some 700 kilometers of the Tisza River and the Danube River were so severely contaminated with cyanide and heavy metals that the governments of Slovakia and Hungary declared a state of emergency. The Tisza disaster became the driving force behind the signing of bilateral treaties that defined areas of responsibility. At the same time, these treaties ensured agreement between neighboring countries on damage regulation, thus exerting the necessary pressure on signatory governments to prevent such disasters.

In the United States, the Colorado River is enduring a similarly dramatic fate. The mighty river that once carved out the Grand Canyon now barely trickles into the Gulf of California. To make matters worse, the household wastewater of several million people, and 600 tons of nitrate, end up in this thin stream every year. The concentration of nitrate is currently four times higher than the legal limit of 50 milligrams per liter.

The Colorado River is subjected to two separate but extremely hazardous sources of contamination. One is a plant in Henderson, Nevada, which has been dumping perchlorate, an additive used in rocket fuel, into the river since the 1950s. Despite several cleanup measures, the plant still releases six tons of this toxic substance into the Colorado every month. The other source, a nuclear waste landfill near Moab, has been shut down in the meantime but its bottom remains unlined. About 12 million tons of radioactive waste have been deposited close to the banks of the river. Experts estimate that 400 cubic meters of contaminated water have leached from this landfill every day for decades, most of it ending up in the Colorado River. For years, the California Department of Water and Health has recorded a constant increase in radioactive contamination of the waters of Lake Havasu, 1,000 kilometers downstream from Moab. Considering the fact that the lake supplies drinking water to 16 million people, this is a disaster.

There is no solution in sight. The Colorado River can only be cleaned up if the Moab nuclear-waste landfill is completely excavated and its dangerous radioactive waste incinerated and vitrified in special facilities. No one knows how many years it will still take for radioactive residues to be washed out from the contaminated sediment.

But the biggest problems are not the spectacular chemical and oil spills that dominate international headlines for a few days, sending politicians into a flurry, only to be forgotten and without consequence soon afterwards. Nor are they the many small accidents reported almost daily on the back pages of newspapers. These incidents are usually limited in scope and can often be controlled or cleaned up effectively.

It is the daily contamination occurring imperceptibly which is far more alarming than these big-time disasters, causing much more water pollution than all the accidents and breakdowns put together, and remaining more or less invisible thanks to the oldest and still most common method of wastewater removal. Polluted wastewater is diluted until concentrations of hazardous substances drop to a level where they are considered harmless or tolerable. Huge quantities of environmentally hazardous and health-threatening chemicals are discharged into rivers in this way, causing lasting damage to the aquatic ecology. Surprisingly, this is common practice even in highly developed nations with strict environmental laws and modern wastewater treatment technology.

Water experts throughout the world agree that this method of wastewater disposal is useless because it does not eliminate toxins but merely distributes them in a larger quantity of water. Subsequent treatment becomes very expensive, and in many cases is practically impossible because of the greatly increased volume of water. The only sensible alternative is to dispose of hazardous substances at the point of source in the industrial facilities using them. In most cases it is easy to precisely locate sources of toxins since they are quite limited in extent, for instance in a specific stage of production in a large chemical plant. Small, specialized treatment plants are well able to filter limited amounts of contaminated water before returning it to the natural water cycle.

For production facilities to have a viable future, they will need to recycle used and polluted water within a closed system. Most facilities can be retrofitted without incurring unreasonable expenses—such measures can even reduce costs in the long term. The Julius Schulte paper factory in Düsseldorf, Germany, developed a technology in 2004 which cleans and softens wastewater so thoroughly that it can be reused in the paper manufacturing process. In this way, the factory saves 260,000 cubic meters of wastewater or 400,000 euros in wastewater disposal fees per year, and no longer needs an effluent pipe (→ pp. 368).

The most sound and sustainable ecological alternative is obvious. Hazardous substances must be replaced by environmentally-friendly ones. Decades ago, alarming levels of pollution in rivers and seas due to detergents led to a ban on phosphates and non-degradable tensides in many countries. These substances were replaced by biodegradable chemicals. Such alternatives exist today for numerous products, and independent process engineers believe that environmentally safe substitutes can be found for many more.

The use of sustainable, non-toxic, and degradable chemicals, and the development of wastewater-free and thus water-efficient industrial manufacturing processes may sound like an ecologist's dream. But this dream is a necessity, especially in developing countries. Rivers are almost the only source of drinking water for the majority of people in Asian and Latin American urban centers. Transforming heavily polluted river water into safe drinking water isn't technically or economically feasible in most cities. Ultimately there is no choice but to take action, knowing that waste does not belong in rivers and seas—even though this disposal practice has been normal procedure for over a century. Alternatives are available and we must put them to use.

Very little is
the effects t
of the tens
of chemical
use have on
and human

known about

at **95 percent**

f thousands

s in daily

plants, animals

beings.

Consumers want to know what resources are used and what kind of pollution is caused in the production of consumer goods. The sale of washing machines in Brooklyn, New York, 2005. Kathy Willens/AP Photo

Is it fun being stingy? The demand for cheap consumer products is accelerating the mass production of goods under dubious conditions. What chemical bouquet will this customer take home today to complement her clothes? Martin Parr/Magnum Photos

Equal rights for all. Nobody would want to deny this Chinese family prosperity and consumer goods. At the same time, there are nowhere near enough raw materials and energy resources to transfer western consumption patterns to billions of consumers. A young couple tests chairs at IKEA in Beijing. Walter Schmitz/Keystone/Bilderberg

Thousands of tons of personal hygiene products are flushed down bathroom drains and into sewage treatment plants. So far, industry has successfully blocked legislation that would compel it to investigate whether this waste is harmful to human beings, animals and the environment. A supermarket in Sao Paulo, Brazil, 2002. Stuart Franklin/Magnum Photos

Shopping consciously: Anyone who simply buys what is on their shopping list misses out on the supermarket's latest bargains–designed to increase sales–and their backpacks full of chemical additives. The "offer highway" in a London supermarket. Photo taken in 2003. Peter Marlow/Magnum Photos

Chemicals policy –sustainable disappointment

The majority of hazardous substances are not released to the environment in direct discharges from factories, but from industrially manufactured products. Foods, textiles, computers, cars, construction materials, drugs, detergents, paints, cosmetics, cell phones, and home furnishings nearly all contain synthetic or chemically-treated materials. Most constituents of synthetic chemicals are harmless or biodegradable and can be disposed of safely, or recycled, or industrially incinerated. Others however are toxic and degrade very slowly, if at all. They continuously accumulate in the environment, and some are even biologically active.

Whether harmless or toxic, these products end up in the trash, in household wastewater, garbage dumps, landfills, and temporary disposal sites. All over the world, they pile up in growing garbage heaps and in the end are released together with their degradation products into the air, the ground, and into water. Governments until now have dealt with this problem in a reactive way, only taking steps and enforcing laws when, after years or decades of ongoing pollution, it has turned out that certain substances are harmful to the environment or human health.

Neither consumers, environmental agencies, nor health authorities know which detergents, softeners, solvents, flame retardants, stabilizers, and other chemicals are contained in tens of thousands of consumer products in everyday use. Usually, the components of consumer products are tested only when there are indications of a threat to the environment or human health. Many years can pass before chemical ingredients or additives are fully tested and suspicions are confirmed or not. Meanwhile, these chemicals often remain in use for years without any restrictions.

An increasing number of environmental scientists rightly believe that using untested substances is irresponsible. Once hazardous substances have begun accumulating in the environment, it can take years or even decades for nature to at least partly recover. PCBs (polychlorinated biphenyls), for instance, are still widely found in the environment although their use has been banned for decades.

Most manufacturers and users of chemical products and a majority of economic policy-makers consider strict controls and environment-friendly regulations to be incompatible with liberal economic systems. They believe such controls are unnecessarily expensive and deter investment, blocking research and innovation. In short, controls get in the way of progress. They argue in favor of self-imposed restrictions and voluntary agreements between government authorities and industry.

This makes it difficult to enforce effective policies on chemicals. Even within the European Union, the economic region with the world's most progressive environmental laws, any policy on chemicals remains incoherent and full of loopholes. Every new substance developed since 1981 and manufactured in larger quantities must be registered together with relevant data and test results so that EU authorities can assess its possible hazards. Altogether 3,000 new substances have been evaluated in recent years.

The problem is that approximately 100,000 chemicals that were on the market before 1981 have, for practical reasons, not been assessed. There is barely any reliable data for 97,000 known "old" substances of which less than 1,000 tons each are produced annually. To date, a mere 80 such old substances have been thoroughly assessed. The EU's original intention to retrospectively evaluate the most important old substances has meanwhile been abandoned.

The EU decided in 1999 to consolidate its policy on chemicals in a new, stringent, and effective regulatory framework directive called the Registration, Evaluation, and Authorization of Chemicals (REACH). Its goal was to protect the competitiveness of European industries on the world market while giving just as much weight to the protection of health and environment, granting these issues clearly more importance than they had ever had before.

The REACH draft published by the European Commission in 2003 called for all substances (old and new), of which more than one ton are annually produced or imported into the EU, to be entered into a central database by 2012, including information on their properties, uses, and risks.

Based on the data submitted upon registration by companies, substances suspected of being harmful to people's health or the environment are to be thoroughly analyzed and evaluated. Experts estimate that this applies to approximately 30,000 chemicals. According to the REACH draft, even a chemical classified as a Very High-Concern Chemical– meaning potentially carcinogenic, bioaccumulative, or responsible for genetic mutation–can remain in use as long as associated risks are "adequately controlled." This vague wording allows plenty of room for interpretation. Moreover, this regulation does not apply to substances for which environment-friendly substitutes are not available and which have socially or economically "justified" uses.

The REACH draft nevertheless contains a trend-setting innovation. It proposes that the burden of proof lie with manufacturers and not with government authorities, clearly in contrast to standard procedure in the past. If adopted, companies will be responsible for carrying out all necessary tests and risk assessments, and covering these costs themselves.

While environmental organizations criticize the vague wording and demand more precise and stringent legislation, industry thinks the REACH draft goes decidedly too far, arguing that such strict regulations put EU industries at a grave competitive disadvantage compared to countries with more lenient legislation. Furthermore, delays caused by the exhaustive testing required by REACH prevent a rapid launch of new products, resulting in additional disadvantages for EU businesses. Industrial lobbyists threatened to relocate factories to other countries not demanding this kind of assessment. They argued that passing costs on to the parties responsible would put an intolerable burden on small and medium-sized enterprises in particular.

After years of fierce debate between representatives of industry and environmental organizations, the EU parliament accepted a compromise proposal in November 2005 which gave more weight to the objections presented by industry. To ease the burden on small and medium-sized businesses, the testing process for a chemical is simplified if annual production of a substance does not exceed 100 tons. This also applies to chemicals that consumers do not come into direct contact with. Contrary to the original objective of REACH legislation, tens of thousands of old chemicals will thus be exempted from standard testing requirements. REACH guidelines are not expected to go into effect before 2007.

The controversy surrounding REACH has set an important precedent for global policy on chemicals. The EU is the most powerful trading power in the world and is responsible for 20 percent of all imports and exports. Its decisions are not without consequence for the rest of the world.

The first steps towards developing a global policy on chemicals were taken in the early 1970s. But twenty-seven years went by before a precise and internationally binding agreement could be reached. This is the Rotterdam Convention on Hazardous Chemicals and Pesticides (Internationally Legally Binding Instrument for the Application of the Prior Informed Consent Procedure for Certain Hazardous Chemicals and Pesticides in International Trade).

The Rotterdam Convention, otherwise known as the PIC Convention, went into effect in February 2004, but is still hardly a significant first step towards an effective global policy on chemicals. It does not stipulate production bans and is limited to the monitoring of trade. Even the chemicals it classifies as especially problematic may be exported if the country of destination is informed of the risks involved and gives its consent (Prior Informed Consent–PIC). The convention currently lists several pesticides and some industrial chemicals, including DDT, PCPs (pentachlorophenol, its salts and esters), and PCBs. There are plans to add every carcinogenic type of asbestos to the list within the next few years.

The Organization for Economic Cooperation and Development (OECD), the World Health Organization, the UN Environmental Program (UNEP), and the International Labor Organization led the way towards formulating a global policy on chemicals in the 1970s. Three UN organizations in

1980 initiated a joint program for chemical safety. But it wasn't until 1992 at the Earth Summit in Rio de Janeiro that the issues outlined in Chapters 17, 18, and 19 of Agenda 21 were brought together for the first time. Agenda 21 linked chemical and health policies to ecology, emphasized the central role of water, and gave sustainability high priority.

Since the Rio conference, a number of initiatives and commissions on chemicals and water policy have been set up:
– Two UN commissions, the Commission on Sustainable Development and the Intergovernmental Forum on Chemical Safety, were founded at the Rio conference. Their purpose is to monitor the national and international implementation of Agenda 21.
– A plan for action was decided on at the Johannesburg World Summit on Sustainable Development in September 2002, "aiming to achieve, by 2020, that chemicals are used and produced in ways that lead to the minimization of significant adverse effects on human health and the environment." The Johannesburg statement, however, is merely a declaration of intent and not a legally binding regulation. In other words, it is a "soft law." The plan's documents do not contain quantitative guidelines, and hazardous effects should simply be considerably lower than at present.
– In September 2003 in Nairobi, UNEP decided to start preparing for a conference to work out a "Strategic Approach to International Chemicals Management" to ensure that all current and future activities linked to chemical safety would be analyzed and coordinated.
– The OECD initiated a program to coordinate the systematic assessment of risks and analyses of substances produced in quantities of more than 1,000 tons (known as HPV or high production-volume substances).
– The Stockholm Convention on Persistent Organic Pollutants (POPs) was adopted in 2001. This global agreement for the first time banned several POPs that are particularly threatening to human health and the environment. The current list bans only eight substances—seven pesticides and the PCBs used in flame retardants, coolants, transformers, and electrical appliances. The Stockholm Convention laid down restrictions on DDT, dioxins, and furanes, some of which form when toxic waste is incinerated. However, a number of exemptions are valid even for banned substances. The POPs Convention was designed to be dynamic, its list gradually expanding in the future.
– These approaches are a long way off from the systematic and internationally coordinated chemicals policy that the EU's REACH program originally intended. For a variety of reasons, powerful industrial groups from highly industrialized nations and their governments as well as several developing countries are strongly resisting a stricter policy on chemicals. Short-term economic and financial interests still have priority over the long-term threat to humans and the environment. Future generations will have to carry the burden.

The World Trade Organization (WTO) has enacted two regulations which could seriously impede or even deter the implementation of regional treaties such as the EU's REACH directive. For instance, WTO members are only permitted to enforce protective measures that do not discriminate against other WTO members. In other words, nations or nation groups such as the EU cannot ban imports that were produced legally in other countries. Furthermore, the WTO has decided that restrictions or prohibitions must be "appropriate," meaning that ecological goals should not obstruct other interests, especially economic ones. These regulations create a number of political and legal loopholes undermining strict regulations in some countries.

Klaus Töpfer, commissioner of UNEP, the UN environmental program, declared that in spite of all the international conferences, "implementation of an integrated management of water resources has not been widely successful in either industrial or developing countries." Problems were still being dealt with "on the basis of fragmentary sectoral approaches." He went on to say that the political will obviously did not exist to regulate ecological, economic, and sociocultural interests on an equal basis. After the 2002 Johannesburg Summit, a German newspaper summarized this criticism in a headline which read: "Summit of sustainable disappointment."

Everything comes back in the end: The desirable, glittering items that filled the shelves yesterday are the disposal problem of today. The recycling of electronic waste products in Regensdorf, Switzerland. Photo taken in 2004.

Walter Bieri/Keystone

The more toxic a product, the more difficult and expensive to dispose of it in an ecological manner, and the more likely its illegal dumping. An unauthorized special waste dump in Great Britain. Robert Brook/Keystone

A rare sight: For each device bought new, an old one ends up on the dump. Scrap fridges are collected at an outdoor site near Lewes in Southern England in an effort to recover coolants hazardous to the environment. Photo taken in 2003. Gerry Penny/Keystone, EPA Photo

More than 21 million VW Beetles alone were produced. Disposing of the early models was an easy matter. Nowadays, however, composite materials and problematic fixtures make it increasingly difficult to recycle new models. A scrapyard in Salzgitter. Keystone/dpa files

Hydropower– a mixed blessing

Hydropower is by far the most important of all renewable energy sources. Hydroelectric power plants produce about one-fifth of the electrical energy consumed around the world. This puts water power in second place after the fossil fuels coal, oil, and natural gas, with nuclear energy coming in third. All other energy sources like sun and wind, added together, make up a much smaller share.

Although climatic zones and conditions in various regions of the world are vastly different, the electricity generated by water power could be increased just about anywhere. Experts believe that the current worldwide capacity of hydroelectric power plants could be doubled or even quadrupled. Emerging countries like China, India, Pakistan, Brazil, and Nigeria, which have large populations and an abundance of water, have not exhausted the potential of their water resources. But hydropower could also be used to significantly relieve the most pressing energy problems facing many poor developing countries.

Since the 1960s, large emerging countries, with the support of the World Bank and of leading industrialized nations and their multinational construction and technology businesses, have concentrated on building huge dam reservoirs and hydroelectric power plants with enormous capacities. Hundreds of thousands of people have been, and will be, resettled to make way for these projects. Although these

major dams are highly controversial for this very reason,
most countries are proceeding with their projects,
and unswervingly continue to plan even more.

The proponents of hydropower argue that it is efficient,
inexpensive, and ecologically harmless since hydroelectric
power plants are much more efficient than comparable
coal-fired or oil-burning plants. Water "fuel" is for free and
hardly produces ecologically harmful greenhouse gases.
But we've known for a long time that hydroelectric power
plants in combination with major dams are far less harmless than claimed. Not only do they destroy the life-support
base of people living in the valleys that are flooded, they
also seriously damage the whole river landscape as far
down as the rivers' estuaries. Previously fertile farmlands
are no longer fertilized by nutrient-rich river sludge, huge
feeding and spawning grounds for fish are deprived of
nourishing sediment, and river deltas are eroded and sink
into the sea. In the end, it's not hundreds of thousands of
people who are affected, but hundreds of millions.

Large hydroelectric power projects are mainly geared
to supply electricity to growing industrial agglomerates and
metropolises. Still, approximately two billion people in
the world have no access at all to electricity, mostly in poor
rural areas. Small and decentralized hydropower stations
would be much more suitable for providing power in many
places than large-scale facilities. Such plants can be
realized without major disruption of the landscape and
could get by using local supply networks. Since they
generate only very limited amounts of power, people would
have to use the available electricity sparingly, particularly in water-scarce regions. This is not necessarily bad,
say energy experts with an eye to the future—unless we
learn to use available resources more efficiently and wisely, we won't be able to solve the energy problems of the
future.

The number of people forced to leave their homes over the past decades to make way for reservoirs adds up to many millions. The 2,300-year-old city of Fengjie was razed to the ground to make way for the water dammed by the Three Gorges Dam in China. Reuters/China Photo

The people who are resettled for hydropower projects lose both their homes and their traditional livelihoods. In May 2003, the residents of Fengjie leave thier destroyed city which will soon be submerged under 135 meters of water.

Reuters/China Photo

A symbol of progress and a tourist attraction for Chinese: The monumental, five-stage locks of the Three Gorges Dam link the reservoir with the lower course of the Yangtse. Keystone/AP

In the past 5
number of da
the world ha
from **6,000**
In the proces
80 million p
been resettl
the entire p
of Germany

0 years, the
ms around
s **increased**
o **45,000.**
s, almost
eople have
d—**more** than
opulation

Endangered livelihoods: When the dams planned on the Mekong are built, the seasonal floods will cease that supply the fertile mud essential for fishing on Cambodia's vast lake, Tonle Sap. John Vink/Magnum Photos

Social life centres on the river: In Vietnam's Mekong Delta, the river shapes every aspect of life, food, transport, seasons and boundaries. Berthold Steinhilber/Keystone/Bilderberg

Too wide and powerful for bridges: The lower course of the Mekong can only be crossed by ferry; a motorcycle taxi in Kaoh Reah Koaom, Cambodia. Andy Eames/Keystone/AP Photo

Fish, vegetables and rice. The people are indebted to the river for nearly all the products at Phung Hiep market in the Mekong Delta. Hiroji Kubota/Magnum Photos

Distrustful of dams: Will the new dam bring prosperity or simply disrupt their traditional lifestyles?
Near Luang Nam Tha, Laos, 2004. Marcus Rhinelander

The true cost
of hydropower

The American inventor Thomas Edison officially opened New York's first power plant on Pearl Street on September 4, 1882. The generator weighed 27 tons and produced a mere 100 kilowatts, or enough electricity to power 1,000 lamps. At that time, nobody would have expected that a few decades later hardly anything in the world would be lit or moved without electricity. Today we no longer measure the world's consumption of electricity in kilowatts, megawatts, or gigawatts, but in terawatts, meaning billions of kilowatts.

Since 1990 alone, global power consumption has increased by almost one-third. The International Energy Agency forecasts that by 2020 demand for electricity will go up another 46 percent from today's consumption of about 15,000, to 22,000 terawatt hours per year. It is largely the rapidly growing demand for energy in highly populated emerging countries such as China and India that is behind this growth. To satisfy the needs of their ambitious industrial development plans, all large emerging nations have expanded their power generation since the 1970s much more rapidly than industrialized nations have. While worldwide production has tripled since the 1970s, in China it has increased twenty-one times, in India eight times, and in South Korea as much as thirty-two times. Turkey today produces fifteen times more electricity than it did in 1970, Brazil thirteen times more, and Mexico six times more. China has become the world's second-largest producer of electricity, while three other de facto emerging nations, Russia, Brazil, and India, are among the top ten.

These figures hardly reflect the reality of people's lives there. China may be the world's second largest producer of electricity, but if figures are calculated on a per capita basis, the country comes in only eighty-fifth, lagging behind poor countries like Armenia, Tunisia, Paraguay, and Zimbabwe. If each person in China used only half as much electricity as the average North American, China would have to increase its current power production thirty-three times. This figure would be a staggering two-and-a-half times more than today's total global output.

As absurd as such calculations may seem at first, they do make one thing clear. Attempting to orient the expansion of power production solely on demand forecasts won't lead to sustainable solutions. Demand forecasts are wishful thinking, only highlighting how much electricity would be needed to meet the desired targets of industrial and corporate development. Social, ecological, and economic costs fail to be considered in these calculations.

Some 61 percent of the energy needed to generate power is supplied today by fossil fuels—coal, natural gas, and oil. Hydropower comes in second at 20 percent, followed by nuclear power at 17 percent. The power sourced from wind and solar energy makes up no more than 2 percent. Experts reckon that these percentages will not significantly change in coming decades. An ominous prediction—fossil fuels are already no longer abundantly available and will become increasingly scarce and more expensive. They are also largely responsible for worldwide air pollution and global warming. It's only logical that hydropower, as the most important renewable energy source, is playing an increasingly significant role in energy experts' scenarios of the future.

Nature has dictated that the availability of hydropower on Earth is very unevenly distributed. In sixty-five countries, mainly in South America, southern parts of Africa, northern Europe, and European Alpine nations, hydropower supplies more than half of total electricity. In thirty-two countries it produces more than 80 percent, and thirteen countries get their electricity almost exclusively from hydroelectric power plants.

But these figures don't say much about how hydropower is really distributed. In the United States, the world's largest producer of electricity, hydropower accounts for only 7 percent of the total, but the U.S. produces 266,000 megawatt hours per year, twice as much as Norway, which relies on hydropower for 99 percent of its electricity. This output is also five times more than that in all eleven countries in southern Africa, which rely on hydroelectric power plants for more than 50 percent of their electricity.

What sparks the imagination of energy experts is the fact that in many places the technically feasible capacity of hydropower has yet to be fully exploited. Depending on how estimates are made, available technology could theoretically allow worldwide hydroelectric power to double or even quadruple, from currently 3.2 million megawatt hours per year to 7 or 14 million megawatt hours per year. More efficient technology in the future could perhaps increase these figures even more.

Experts at the International Hydropower Association believe that many developing and emerging nations in particular have good potential for hydropower development. While Europe and the United States already use more than 50 percent of their economically viable resources, many developing nations exploit only 20 to 40 percent of their technically feasible hydro potential. Experts at the World Energy Council are even more optimistic, calculating that Africa, where hydroelectric power plants today generate 70,000 megawatt hours per year, has an additional potential of 1.9 million megawatt hours—twenty-seven times the amount of electricity it produces today! They also reckon that Asia, currently producing 330,000 megawatt hours of hydroelectric power, has an unexploited potential of 4.9 million megawatt hours, and that in Latin America, this potential is 2.7 million megawatt hours.

Indeed, endeavors to exploit water resources to the technical limit have produced amazing results. The twenty water turbines at the Itaipu hydroelectric power plant on the Paraná River in the border region between Brazil and Paraguay have a total capacity of 14,000 megawatts. This covers 78 percent of Paraguay's total power needs, and 25 percent of Brazil's, a country of 180 million people. The Guri hydroelectric power plant in Venezuela generates two-thirds of the electricity for the 25 million people living in Venezuela. In Turkey, nineteen hydroelectric power plants in the Southeast Anatolian project will increase the country's electricity production by more than one-fourth, once the project is completed. In the long run, the Turkish government wants to more than quadruple its hydroelectric power plants from today's 125 to 564, which will more than double Turkey's current power production. In Canada, the world's leading producer of hydroelectric power, 450 hydro plants supply 60 percent of Canada's annual power output. Canada's department of energy believes that if all technically exploitable resources were used, hydroelectric power capacity could be tripled.

Hydropower is an important future option for China, already the world's second largest producer of electricity. The world's most populated country, now experiencing an explosive industrial boom with no end in sight, today still generates almost three-fourths of its electrical power from fossil fuels, burning far more than one billion tons of hard coal a year (and this figure is rapidly increasing), which is one-third of total world production. China emits 3.4 billion tons of CO_2 each year, making it Earth's second largest atmospheric polluter after the United States.

The Chinese government assumes that demand will increase by 24,000 megawatts each year. To meet these enormous energy needs, there are plans to build large coal-fired power plants year for year in the future. But for ecological reasons, China is seeking to substantially lower the share of electricity generated by fossil fuels. By 2020, to compensate for this decrease, the share of nuclear energy is to increase from 1.5 to 4 percent, and of hydropower from today's 24 to 33 percent. As modest as these figures may seem, China would have to put two or three nuclear reactors into operation every year to meet these goals, as well as nearly double the capacity of hydroelectric power plants in the next fifteen years from 95,000 to 175,000 megawatts. In other words, every four years, large-scale facilities with a total capacity equaling that of the controversial Three Gorges Project, the world's largest hydroelectric power plant, would have to be added to the grid. In the long run, experts hope for even more—China's technically usable hydropower reserves are supposed to range between 380,000 and 675,000 megawatts.

If it were a question of technical and economic interests only, the unlimited exploitation of hydropower would be a smart option. Water has by far the highest potential of all renewable resources—wind and solar energy can't generate nearly as much power. The energy potential of flowing water can be transformed into electricity with a unique yield of more than 90 percent. In comparison, the electricity output of coal-fired power plants and light water nuclear reactors is much lower, yielding respectively no more than 50 percent and 30 percent of the primary energy used. Finally, even though the construction costs of large dams and hydroelectric power plants are enormously high, their long life spans of eighty to one hundred years, their low operating costs, and the fact that water itself is free, mean that the production costs of hydropower are far lower than the costs of generating power from coal, natural gas, or nuclear energy. (Experts don't always agree on this last point. World Dam Commission studies have revealed that hardly any of the large hydroelectric power plants achieve the capacities calculated in advance, that maintenance costs are much higher than predicted, and that dam reservoirs lose volume because they fill up with gravel and sediment much more quickly than expected, lowering power capacity.)

It is often argued that hydropower is also clearly superior to fossil fuels from an environmental standpoint since it produces thirty to sixty times fewer greenhouse gases than coal or oil-fired power plants and has practically no net effect on the climate. Even better, hydroelectric power doesn't directly cause water quality to deteriorate.

These characteristics easily obscure the fact that the ecological, social, and wider economic consequences of large-scale hydropower facilities may be devastating. Dam reservoirs inundate ecosystems and populated areas, and large-scale dams drastically change the entire lower reaches of rivers, right down to their estuaries. In the case of large rivers, the effects can often be felt for thousands of kilometers downstream.

A case in point is a project that China is realizing on the upper reaches of the Mekong River. (→ p. 507, pp. 365) It will have dire consequences for hundreds of millions of people in China, Burma, Thailand, Laos, Cambodia, and Vietnam, fundamentally changing the character of landscapes all along the lower Mekong. China is planning to build a series of eight large dams and hydroelectric power plants along a stretch of less than 800 kilometers of the river. This project is expected to have a total capacity of around 15,000 megawatts. The plan also calls for making the Mekong navigable year-round for larger riverboats along a stretch of more than 880 kilometers between the cities of Simao in Yunnan province and Luang Prabang in Laos. To date, two of the eight dams have already been built, one is to be completed by 2010, two more are to be put into operation by 2017, and the last three dams are still in the planning stages.

Up to 100,000 people will have to be resettled, and many towns and villages along the river will sink into the reservoirs, as will an archaic, still unspoiled river landscape with spectacular cataracts and waterfalls. The government in Beijing is trying at all costs to silence the growing number of Chinese critics of this mammoth project in order to prevent people from suggesting alternatives. As in many similar cases, this project is meant to serve not only the common good, but also particular interests, be they foreign capital investors, local district administrators, or managers of state-owned or privatized construction and financial consortiums. An interesting little detail—more than 50 percent of the enterprise responsible for planning and constructing the Mekong Project belongs to the Yunnan Huaneng Lancanjiang hydropower company, itself a subsidiary of the globally active Yunnan Huaneng development company, whose president is Li Xiaopeng. He is no other than the son of Li Peng, who until recently was the second most powerful man in the

Chinese party hierarchy, himself a hydraulic engineer and the driving force behind the Three Gorges Dam.

But the construction of these dams on the Mekong will do away with a natural phenomenon which has shaped the livelihoods of people along the river for thousands of years–the seasonal flooding of large areas along its lower reaches. At the height of the rainy season, no less than one-third of the land surface in Cambodia, for instance, is covered with water. Flooding is largely responsible for the lower Mekong and its enormous delta being among the world's most fertile lands. Chinese dams upstream will however hold back water in the rainy months and thus regulate river flow, ensuring a constant output of electricity and leaving the Mekong with a constant water level almost all year. This will put an end to the annual flooding which, in a natural way, has irrigated and fertilized rice paddies with river sediment from Myanmar to Vietnam. The regulation of the Mekong will affect the unusually fertile "gardens" –broad river banks and river islands in Cambodia, Thailand, and Laos– that surface from the river when seasons change, allowing farmers to grow an abundance of vegetables, fruits, mushrooms, berries, and soybeans. These areas contribute a substantial share to agricultural production. Some 90 percent of Laos' cultivated areas are dependent on the Mekong's waters, in Cambodia this figure is about 70 percent, and in Vietnam 50 percent. Hardly anyone dares to predict what consequences the regulation of the Mekong's flow will have on these cultivated areas, which supply more than 100 million people with food.

The rich fishing grounds of the Mekong and the coastal waters near its delta, where the spawning grounds of many saltwater fish are located, are also endangered. Each year, fishermen catch more than one million tons of fish in the Mekong, no less than the yield of saltwater fish nourished by sediments washed into the sea by the river. In Cambodia, fish provides 80 percent of the animal protein in the diet. In Cambodia, Laos, and Vietnam, the livelihoods of several million people depend almost entirely on fishing. Finally, the huge Mekong delta itself is in danger. Floodplains in river estuaries suffer from erosion unless they are replenished by sediment each year. The same changes would shape the Mekong delta as they have the Nile delta since the construction of the Aswan dam in 1971. Since seasonal floods and their accompanying sediment have been held back by the dam, the Nile delta coastline has receded several dozen meters each year.

In short, the expansion of China's hydroelectric power production on the upper reaches of the Mekong will have enormous consequences for the living conditions of several hundreds of millions of people, even though they live more than 1,000 kilometers downstream, far away from China. Added to the tremendous subsequent social and economic costs, the extent of which no one can possibly guess at this point, there will be incalculable ecological cost incurred by erosion, the destruction of wetlands, forests, and river banks, and the loss of diversity in animal and plant life. Finally, economic estimates aside, the Mekong project will destroy a magnificent and irreplaceable natural landscape, unique on this planet.

A similar fate threatens the Salween, Southeast Asia's second longest river, which flows through China's Yunnan province only 50 kilometers away from the Mekong, its course parallel to it for hundreds of kilometers. It crosses several provinces in Myanmar (Burma), partly forms a boundary with Thailand, and finally empties into the Indian Ocean at the Burmese harbor town of Mawlamyine. China intends to build a cascade of thirteen major dams along a little less than 500 kilometers of its upper reaches. Their total capacity of 21,000 megawatts is supposed to surpass that of the Three Gorges Project. Here too, the Chinese government is proceeding with its plans without heeding the criticism voiced by the affected population and downstream states, not to mention taking ecological criteria into consideration.

As if this weren't enough, Myanmar and Thailand farther downstream want to rigorously transform the Salween, one of Asia's last unspoiled waterways, into a pure power supplier. Up to five major dams with a total capacity of 10,000 megawatts are to be built in coming decades. In December 2005, government representatives of Thailand and Myanmar signed an agreement on the first project, the Hat Gyi Dam, which will be located in the border region between the two countries. A win-win situation for both, said Kraisi Kanasuta, president of Egat, Thailand's semi-private national electricity corporation. Since Thailand's domestic dam projects are highly controversial and face vehement opposition from the population, Thailand is interested in buying cheap electricity from abroad to meet its rapidly growing demand, without having to fear opposition. In return, Myanmar, one of the world's poorest countries, gets badly needed foreign currency. Hydropower from another four dams planned by the Burmese military dictatorship, internationally isolated due to its serious human-rights violations, are supposed to turn the country into Asia's power plant. Myanmar hopes one day to feed up to 10,000 megawatts into the planned transnational grid of the Association of Southeast Asian Nations (ASEAN) through extensive overland power lines, mainly to Thailand.

But Myanmar's hydropower plans are not a win-win situation for the Salween and the people who live there. As with the Mekong, the Salween is the only means of subsistence for millions of farmers and fishermen. The fertility of cultivated areas also largely depends on seasonal flood cycles, and dam reservoirs will also drastically reduce fish stocks. Like the Mekong, the Salween is among the world's richest natural habitats for animal and plant life.

The Burmese military government's hydropower plans will severely affect several ethnic minorities—the Shan, Pao, Lahu, Resu, Wa, Aka, Karenni, and Mon peoples—all of whom live on the Salween. They have all been fighting for autonomy for a long time, which is why they have been discriminated against, deported by the hundreds of thousands, and exiled to Thailand by the military regime since the mid-1990s. If the dam projects go through, many of these peoples will lose their homelands along with their means of subsistence.

The consequences of constructing large dams and hydroelectric power plants are not always as dramatic as for the Mekong and Salween rivers, and rarely are so many people directly affected. But in most places where large hydroelectric power plants are being built, their effects on nature and people in the greater surroundings are much more serious than planners will admit. As (relatively) clean and cheap as hydropower may be compared to coal and natural gas, its economic costs, as well as its incalculable and grave social and ecological consequences, are exponentially higher than for any other form of energy. If rivers are re-engineered solely to generate electricity, they lose their most vital features—their seasonally changing water levels, their cargo of fertile sediment and the abundance of their fish stocks.

In view of these enormous consequences, even the most optimistic forecasts on the yet untapped potential of hydropower can't obscure the fact that mankind will have to make do with much lower rates of growth in coming decades. The obvious solution to this dilemma is to use available energy sparingly and more efficiently, thereby massively cutting back on additional demand. Experts at the Asian Development Bank have calculated that developing countries could lower their electricity demand by about 40 percent just by modernizing their out-of-date power plants and industrial facilities and equipping them instead with the technology long used in developed industrialized nations.

Industrialized nations could also lower their demand for power by 20 percent if they consistently applied already existing technology. There are still many ways to increase the efficiency of machines, electrically-powered motors and household appliances, or to save energy in all steps of the energy utilization chain and in all areas of application, whether industrial manufacturing processes, the railways, households, improved insulation, lighting, combined heat and power generation, the utilization of residual heat, recycling, or the use of innovative materials requiring less energy for their production. Yet the development of energy-efficient technology is still in its first stages because energy has been so cheap for so long.

Even more energy could be saved if we stopped using unnecessary and short-lived products, reduced the mass transportation of goods, and finally if we reorganized daily life and the economy to make sure energy-intensive facilities were used mostly in common. Novatlantis, a project run by several institutes of the ETH domain in Switzerland, indicates that the per capita consumption of energy in industrialized nations could be reduced by two-thirds in the next fifty years without any significant reduction in the standard of living, even with income and consumption levels continuously climbing at the current rate. Scientists are convinced that the consumption of electricity could be accordingly reduced in a "2,000-watt society." Even if this vision seems utopian since it does not correspond to free-market logic and necessitates significant directional state intervention, it does show that limiting worldwide energy consumption to today's average level, which indeed does amount to 2,000 watts per capita worldwide, is at least technically realizable.

Light on, light off: The development of small hydroelectric power plants could improve and stabilize the power supply in Cuba, too. Blackout in Baracoa. David Alan Harvey/Magnum Photos

Light on, light off: In western industrial societies, there are plenty of ways of using energy more efficiently. Christmas lighting in Bahnhofstrasse in Zurich. Alessandro della Bella/Keystone

About **180,0**
electricity co
around the g
and mini hy
This corresp
the capacity
power statio
quarter of t
being used a

00 megawatts
uld be produced
obe using small
ropower plants.
onds to
of **180 nuclear**
ns. **Only a**
is potential is
t present.

Small hydroelectric power plants are particularly effective in mountainous regions: the channel supplying the microhydro plant at Garam Chasma in Hindukush, Pakistan, a few kilometers from the Afghan border. Photo taken in 2004. Martin Wright/Ashden Awards

Small mountain streams are not only useful for producing electricity, they also continue to operate millstones, fulling mills and hammers. A mill in Nepal. Hahn/laif

Electrical power from small and mini hydroelectric plants opens new horizons. Chalan, Cajamarca, Peru. Picture taken in 1995. Steve Fisher/Practical Action

Small is beautiful

Industry, trade, services, and transportation around the world altogether use approximately four times the amount of electricity that private households do. For statisticians a mere fact, but one that gives power providers a real headache. The cost involved in connecting individual households is huge compared to the electricity they consume. While large industrial power consumers need only a single power line, households require a large, finely branched out, and correspondingly expensive electricity grid.

Wealthy northern countries with an ample supply of electricity, which have continuously developed their power grids for more than a hundred years, are easily able to increase power production and expand their grids at the same time. Both large and small consumers are willing and able to pay for this luxurious service.

This is not the case in developing and emerging nations that are already overtaxed by just producing enough electricity and adjusting output to rapidly growing demand. If they wanted to connect every household, they would have to invest enormous additional amounts of money into the expansion of power grids.

Limited funds, the dominant influence of foreign investors, and the ambitious development visions of emerging nations have clearly shaped their priorities—for them, the expansion of power capacity for industrial and service sector businesses comes first in the belief that this is the engine behind rapid industrialization. Coal-fired, nuclear, and hydroelectric power plants are meant primarily to supply power to large industrial areas and international metropolises. These plans neglect the rural regions that cover the largest expanses in these countries. They also neglect smaller towns far away from large overland power lines, and even more so, city slums. Linking up these places is an unprofitable business since many of the end users, poor farmers, little villages, and slum inhabitants, aren't able to help finance the cost of getting connected to the grid.

The consequences of focusing on industrial needs are far-reaching. One billion people around the world still don't have access to electricity. Most of them live in developing and emerging nations. A study of countries in southern Africa revealed that in thirteen of the twenty-two countries looked at, not even 10 percent of households were connected to the grid. The inequality between rich and poor continues on the national level too. In nineteen of the countries studied, the poorer half of households had to make do with almost no electricity, while more than 90 percent of the wealthiest households (the top 25 percent) were connected to the grid.

To improve this situation, many developing countries rely on the construction of small decentralized hydroelectric power plants. In sparsely populated rural areas, small amounts of electricity usually suffice to cover the local population's most basic needs for lighting and household appliances. In Vietnam, for instance, 130,000 small and very small water turbines generate electricity for more than two million households. China taps the power of its rivers not only for large-scale projects, but also for approximately 60,000 smaller hydroelectric power plants that supply electricity to roughly 300 million people. These are usually much cheaper and more efficient than large-scale facilities.

In Nepal, where 85 percent of the population lives in rural areas, only 13 percent of households are connected to the grid, nearly all of them in urban areas. According to official plans, at least 30 percent of all households will be connected by 2020. While the Terai, Nepal's southern lowlands, can be largely supplied by extending the national power grid, it is almost impossible to connect scattered villages and settlements in remote mountainous regions of the north. In these regions, the focus has wisely been on decentralized solutions such as small hydropower plants and very local power grids. Since the 1970s, a number of local construction and metal-working businesses have been set up with the help of the government and international aid organizations. These businesses are capable of domestically producing all needed components except turbines.

Recently, the World Bank, the United Nations Development Program, and regional development banks have subsidized the construction of small decentralized hydroelectric power plants because they can be tailored to local conditions. River power plants with a capacity of 10 megawatts can supply a smaller town. Micro power plants with an output of one megawatt, powered by smaller rivers, can supply about 1,500 households and small businesses with electricity. Pico hydro systems, cheap and very small turbines with a capacity of just a few kilowatts, are hardly larger than a car engine and can even be installed on small creeks or irrigation canals. They generate enough electricity to supply an individual farm or a small business.

Until the middle of the last century, small hydroelectric power plants were among the most important sources of electricity in many developed industrialized countries. In the small country of Switzerland alone, almost 7,000 small water turbines were in operation in 1924. But since the 1950s, they have gone somewhat out of fashion in most industrialized nations. Many have been shut down or not expanded as originally planned because their aging equipment meant they were no longer competitive. Coal was cheap and nobody worried about greenhouse gases. Soon coal-fired power plants and large storage power stations were supplying power more cheaply than small hydroelectric power stations.

This could quickly change if fossil fuels suffered from permanent shortages and price hikes. In the European Union (EU) countries of western Europe alone, power production could go up every year by 24 terawatt hours if disused plants were rehabilitated and operations restarted with improved technology, and if plants were expanded as once planned. The EU Commission predicted in its 1997 White Paper that power production in small hydroelectric power plants could increase from 40 to about 55 terawatt hours by 2010 (if the economy showed positive development), and could reach 60 terawatt hours between 2020 and 2030. The decisive factor in each case will be whether expansion could be achieved without having significant adverse effects on the ecological quality of the river being used. A small hydroelectric power plant requiring a dam can indeed prevent the upstream migration of fish and crawfish. But the goal of the EU Water Framework Directive is to restore continuous passage for river fauna in Europe's rivers as far as possible by 2015.

Experts estimate that the worldwide potential of micro and pico power plants is about 180,000 megawatts, which equals the output of 180 nuclear power plants. But only 26 percent of that potential, or 47,000 megawatts, is being generated today, half of it in developing and emerging countries. The frontrunner is China with 31,000 megawatts, but India, Brazil, Peru, Malaysia, and Pakistan also have larger shares.

India's Ministry of Non-Conventional Energy Sources estimates that the country's small power plant potential is 15,000 megawatts, of which currently only 1,700 megawatts are exploited. Some 170 smaller power plants are under construction, which will increase capacity by 500 megawatts. But even so, not even one-fifth of India's small hydro potential is being used.

Uganda, one of the world's poorest countries, also wants to push the construction of small and very small hydroelectric power plants. Although water is abundantly available in most parts of the country, rural areas remain practically unconnected. To increase the capacity of small power plants from 17 megawatts today to a possible 200 megawatts in the future, Uganda wants to take advantage of the emissions trading deal provided for in the Kyoto Protocol. By virtue of the Clean Development Mechanism, industrialized nations can compensate for their greenhouse gas emissions by helping other countries to build clean production sites.

Small and very small power plants won't be able to cover a large share of the future power needs of developing and emerging countries, but in rural areas they can supply electricity to millions of households, offering many people a valuable alternative to migrating to big cities. These plants have advantages over large-scale facilities which can't be ignored. Their construction does not require millions or billions of dollars in loans from the World Bank or foreign investors. With some help, small towns, even villages and cooperatives, can independently finance their power supplies. Local businesses can do a large part of the construction work themselves. Maintenance requires few or no trained specialists. Last but not least, small hydroelectric power plants can be designed and completed within a few years, and their long life spans of more than fifty years means they produce relatively cheap electricity. They don't need expensive high-voltage power lines over long distances, and they make do with small, simple distribution networks. Larger intrusions into the landscape are usually unnecessary. At best, these plants make use of natural conditions like waterfalls, rapids, and the rapid flow of steep rivers. Careful and conservative planning keeps negative ecological impact on rivers to a minimum.

Nevertheless, small and very small hydroelectric power plants aren't a general alternative to other ways of generating electricity. The amount of power they produce is far from enough to cover the energy needs of large industrial complexes or big cities. If demand increases, a river's limited resources can't be expanded easily in the way coal-fired and nuclear power plants can.

Patent remedies that are socially responsible and ecologically sound don't exist for covering the growing demand for electricity in coming decades at low cost. The modern vision of being able to control nature through technology, providing everlasting progress and a better life for all, has proven to be a deceptive illusion regarding energy supply. Large-scale technology for generating power in coal-fired, nuclear, and huge hydroelectric power plants destroys the social and ecological conditions necessary for life—far more extensively and permanently than expected. Sustainable energy policy will have to be shaped by two goals. First, the realization of projects should not be reduced to the question of whether such projects are technically feasible or can be funded. The determining factors are whether they really serve the needs of the people concerned for survival, whether they are adapted to local social and natural conditions, and whether they ensure the protection of humans and nature in a sustainable way. Second, forecasted demand can't be the only standard. Sustainable energy policy must be, at the same time, a policy for saving energy—our largest untapped resource is energy efficiency.

Even the most elegant technology has its drawbacks: Large dams disrupt natural water cycles and fundamentally alter the river's eco-systems. The Emosson dam, Switzerland. Photo taken in 2003.

Olivier Maire/Keystone

Whether huge dams are viewed as a curse or a blessing depends on perspective. Devout Hindus protest against damming-up of the Baghirati River by the Tehri dam. As a result, their holy river, the Ganges, has lost one of its most important sources. Allahabad, India, 2006. Rajesh Kumar Singh/AP Photo

Close-ups

GrupoNueva—business to combat poverty
Save the Mekong!
A farewell to wastewater—producing paper without polluting rivers
Water management tests the limits—the case of Oman
Water from fog
When dams get old: Dam Removal in Western North America
The salt of life—oral rehydration therapy
Solar disinfection of water—drinking water in six hours
Novaquatis—a building block for innovative sanitation technology
Solving sanitation problems in Kumasi, Ghana
Hope for Goulburn
The privatization of water utilities in England—not a success story
Is water a prime commodity for investment?
Bali's rice farmers and their water priests
The women of Plachimada
Beyond economy—the Irish experience with water pricing
Of moon, wood, and water
Water, power, and NGOs
The Bangladesh sanitation miracle
An icy divide
New Orleans—a disaster waiting to happen
New Orleans still at risk

GrupoNueva's philosophy is motivated by ethics and philanthropy. It links market and competitive elements to sensitivity, forward-looking plans, human integrity, and care for the plight of poor population groups. Sadly, the purely profit-oriented thinking of many multinational businesses is in practice a long way off from this idea.

GrupoNueva–business to combat poverty

There is a lot of water in Guatemala, but farmers there can hardly use this resource in an optimal way. They are in need of more efficient drip irrigation to increase production, but haven't been able to afford this technology–until now, at any rate.

Amanco, a subsidiary of the Latin American GrupoNueva group, since 2003 has been selling to smallholders (campesinos) in Guatemala a simple and inexpensive irrigation system called 4x4. It regulates the moistening of plant roots, thereby increasing the yield of every harvest by one fifth. This more efficient use means water is available to farmers during four seasons, allowing them to harvest crops four times a year instead of twice (4x4). During the very first year, use of the 4x4 irrigation system typically allowed a smallholder with one hectare of land to increase his income by 230 percent, from 3,840 dollars to 8,832 dollars. This additional money gave the farmer access to public health services and enabled him to send his children to school.

Nevertheless, campesinos are poor. Because they lack regular income, they are given loans only at horrific rates of interest, generally around 30 percent. To be able to sell 4x4 systems at all, Amanco developed a strategy specifically designed for the purpose. First it organized low-interest loans and then persuaded Guatemala's ministry of agriculture to provide financial and technical aid for the 4x4 project. Finally, two local NGOs were found that are now helping farmers find new sales channels for marketing their products.

It is obvious that Amanco is not an ordinary industrial business. But it is banking on increasing sales and does want to make a profit. Hundreds of campesinos have declared their interest in the 4x4 system, so in the next three to five years, Amanco is looking to sell 5 million dollars' worth of micro-irrigation technology in Guatemala alone. But behind the group's business practice is a developmental interest–indeed, a vision.

The trail leads to Stephan Schmidheiny, who was the sole owner of GrupoNueva until 2002. GrupoNueva, a leading industrial group built up and expanded for years by Schmidheiny in seventeen Latin American countries, produces and sells water pipes and irrigation systems, or, more precisely, sustainably produced wood products and building material, through its subsidiaries Amanco and Masisa. At the same time he was involved in these business activities, Schmidheiny also financed the non-profit foundation AVINA, which works together with leaders of civil society and the business community, supporting and providing training for their initiatives towards sustainable development, and building and financing networks among AVINA partners from different sectors of society.

In 2002, Stephan Schmidheiny decided to knit his commercial and philanthropic commitments in Latin America even more closely together. He brought GrupoNueva into a new foundation called Viva Trust (Vision y valores) and at the same time withdrew from management of the holding and the AVINA foundation. This gift–with an estimated value of one billion dollars–means that GrupoNueva's entire economic surplus can be used to benefit charitable causes.

At the same time, Schmidheiny inspired GrupoNueva's managers to set a business course that is ecologically and socially beneficial. One main concern is to do fair and sustainable business with poor population groups modeled on the experience with the 4x4 irrigation system. On the strength of this enterprising developmental approach, Viva Trust hopes to improve the living conditions of Latin America's poorest people– the plan is for thousands of farmers to receive affordable loans with the help of the Inter-American Development Bank. A desirable side effect is that whenever 4x4 technology for micro-irrigation is sold, the profit from sales will indirectly benefit Grupo Nueva. As long as this profit is used in turn for the public good, this approach is beyond all criticism on ethical grounds.

It remains to be seen whether other multinational groups will apply the same strategy. At any rate, GrupoNueva's commitment does indicate to other industrialists that poor population groups in society can also be worthwhile business customers. It would certainly be welcomed if Procter & Gamble, Suez, or the Deutsche Bank were to pursue similar strategies. But as long as profits from their activities must be exclusively paid out as dividends to stockholders, corporations will have to put up with the accusation that they conduct business at the cost of the very poorest.

GrupoNueva's philosophy is motivated by ethics and philanthropy. It links market and competitive elements to sensitivity, forward-looking plans, human integrity and care for the plight of poor population groups. Sadly, the purely profit-oriented thinking of many multinational businesses is in practice a long way off from this idea.

Photo
Chimaltenango, Guatemala. Douglas Marroquín.

It is one of the longest and most powerful rivers in the world, crossing three different climate zones, and providing subsistence for 70 million people. Until now, nature and history have to a large extent protected the Mekong from the effects of civilization. A one-man enterprise is fighting to preserve one of the most diverse and magnificent river landscapes in the world.

Save the Mekong!

The river has many faces and just as many names—the Chinese who live in the precipitous, inhospitable mountain valleys of its upper reaches call it the Lancang Jiang, meaning wild river. Farmers and fishermen in Thailand and Laos call it Mae Nam Kong, the mother of water, because its abundant fishing grounds and fertile riverbank fields make it their most important life-support base. In Cambodia, it's called Tonle Thom, the big water, where its annual floods at the

same time mean life and death, plentiful harvests and destroyed villages. The Vietnamese who live in the huge delta where the river branches out into numerous smaller courses call it Sông Cuu Long, the river of nine dragons.

But for the rest of the world, the most appropriate name for the Mekong, especially its upper reaches, would be "unknown" river. It wasn't until 1994 that Michel Peissel, a

French ethnologist and documentary filmmaker, first ventured into the region of its headwaters. His adventurous expedition advanced into one of the most remote areas of the world, the old Tibetan kingdom of Nangchen, through rocky canyons and windy wastelands, past ruined monasteries, isolated settlements and lonely Chinese garrisons. On September 17, 1994, Peissel and his native guides, nomadic Khampa herdsmen, reached a little mountain stream which bubbled out of scree below the Rupsa pass, 4,975 meters above sea level, and assumed they had found the source of the Mekong. In the same year, Japanese explorer Masayuki Kitamura located a different source dozens of kilometers to the east at the base of the Lasagongma glacier, 5,224 meters above sea level. Several scientific study trips were made to both sources, and the Chinese academy of sciences even made a third suggestion before in 1999 the Lasagongma glacier was officially declared the true source of the Mekong.

The Mekong's headwaters descend between rugged mountains and icy wastelands on the Tibetan plateau and drop steeply down the rocky valleys of the Kham until 400 kilometers downstream, at an altitude of 3,000 meters, they reach the first sizeable city with several tens of thousands of inhabitants. Hardly noticed by the rest of the world, it was here in Chamdo that the fate of Tibet was decided in the fall of 1950 when the Chinese People's Liberation Army gained access to Lhasa after a dramatic battle. The region around Chamdo, home to tenacious Khampa rebels, remained the center of Tibetan resistance for a long time after the Chinese Cultural Revolution.

The Mekong remains an unknown even south of Chamdo, where its waters run parallel to those of the Yangtze and Salween River, only a few dozen kilometers apart, all of them tumbling through deeply furrowed, narrow mountain valleys and wild impassable canyons between mountains as high as 5,000 and 6,000 meters in altitude. Many parts of this region, home to numerous mountain tribes, are areas where access is prohibited by the military. Even today, Yunnan province is one of the poorest regions in the world. It wasn't until the 1980s that the Chinese government decided to develop its remote back territory. After all, the southwest Silk Road went through this region as long as 2000 years ago, linking China to India. Even today, the main road across the Yongbao bridge is still China's most important link to Myanmar (formerly Burma), and to India just less than 300 kilometers away.

Much of the electricity needed to develop industry in Yunnan is to come from a whole cascade of large-scale dams driving eight hydroelectric power plants. The first of these, the Manwan Dam, was completed in 1994, followed by the Dachaoshan dam in 2003. A third, the Xiaowan dam, is under construction and will be finished in 2013. At a height of 300 meters, its dam wall will be one of the largest in the world. The eight Chinese dam reservoirs alone will destroy the natural course of the Mekong along a stretch of more than 600 kilometers, 13 percent of its total length. There will be even more massive intervention when China uses excavators and dynamite to continue its project to make the Mekong navigable for large freighters all year long. Numerous reefs and rapids, today still spawning grounds for many species of fish, will have to make way for the water highway that is supposed to link China with Laos, Thailand, Cambodia, and Vietnam.

Regulating the Mekong with dams and reservoirs will not only profoundly change many diverse river landscapes, it will also bring to a standstill a natural irrigation system which has been in use for thousands of years, driven alone by the change in seasons. This will have far-reaching impact on millions of people. Annual flooding supplies broad valleys with abundant nutrient-rich sediments, making the lower reaches and the delta of the Mekong one of the most fertile regions on Earth. No one knows exactly to what degree, in the long run, yield will decrease on the vast expanses of rice fields and areas cultivated for vegetables, fruits, and soybeans. The only thing for certain is that millions of people will have less food on their plates.

The same holds true for fisheries. More than 1 million tons of fish are caught each year in the Mekong. Regulating the river will destroy numerous spawning grounds and many of the 1,300 fish species making the Mekong one of the most biologically diverse rivers in the world will disappear or be seriously decimated.

Regulation will have a particularly devastating effect on Cambodia. Close to its capital city, Phnom Penh, is the great Mekong lake Tonle Sap, a natural phenomenon unique in the world. Every year in June when the Mekong is flooded, the river that joins the lake with the Mekong reverses its direction of flow. For four months, water flows "uphill" from the Mekong into Tonle Sap. The lake grows to five times its normal size and covers about one-seventh of Cambodia's total area. The enormous clearcutting of forests in the region has already massively changed the ecology and hydrology of Tonle Sap—its water level drops more each year, causing the ever shallower lake to warm up so much that many fish species can no longer survive in it. Fishermen still catch about 230,000 tons of fish in Tonle Sap each year, providing a staple food for two to three million Cambodians. If the Mekong is regulated by dams, the seasonal floods feeding the lake will fail to appear, and most of Tonle Sap will silt up within a few decades. This will have enormous consequences for fish stocks and extended rice cultivation in flooded zones along its shores.

But it's not just regulation that will dramatically affect the sixty million people living near the Mekong. Rapidly enforced industrialization, especially in China, will also take its toll. This can best be observed where the Xier tributary flows into the Mekong. The Xier's waters flow from Lake Erhai. On the lake's southern shore lies the district capital of Xiaguan, with some 500,000 inhabitants and many large paper, chemical, and food-processing factories. Almost all wastewater from the city, its factories, and from surrounding villages flows untreated into the formerly pristine lake, transforming the Xier into a dark-brown, stinking, foamy wastewater canal. Massive river pollution like the Xier's is still rather the exception—but if industrialization along the Mekong and its numerous tributaries continues as intensely as the Chinese government intends, the Mekong will be just as badly polluted in a few decades as the Yangtze already is today, its waters useless.

This development must be of concern to the five downriver countries of Myanmar, Laos, Thailand, Cambodia, and Vietnam, whose populations are largely dependent on the Mekong's waters. Until now, China has strictly refused to discuss and agree its plans with countries downstream. The situation is further complicated by the fact that China and Myanmar will not accept open invitations to join the Mekong River Commission (MRC) founded in 1995 by Thailand, Laos, Cambodia, and Vietnam to facilitate cooperation on the sustainable development of a shared natural resource.

The MRC has two communist states as members and boasts an open policy of basin-wide cooperation. Thus its agenda differs broadly from that of its predecessor, the Mekong Committee, which was founded in 1957 at the initiative of the United States with the intention of using it as an economic and political instrument to curb the spread of communism in the region. It never sought to include China and Myanmar.

Australian Mick O'Shea, who runs the Mekong First Descent Project, has taken a personal interest in creating awareness of the impending destruction of the Mekong, currently progressing without receiving much global attention. An adventurer and advisor on eco-tourism, he traveled repeatedly throughout the Mekong River region and for six years reconnoitered and documented sections of the river, finally traveling by canoe from source to mouth in an extremely arduous five-month-long expedition. Altogether his team has taken 15,000 pictures and shot more than 100 hours of film. Overpowered by the unique beauty of this river landscape and its natural and cultural diversity, fascinated by its thousands of years of history, still to be witnessed here and there, impressed by the people who deal with their river within a unique symbiosis, and shocked by the threatening destruction of such irretrievable natural and cultural wealth, Mick O'Shea has initiated a one-man enterprise to save the Mekong.

His enterprise is not just a comprehensive documentary effort to make a record of the Mekong from its source to its mouth in books, films, and the Internet–it also involves setting up a foundation that sees itself as a voice for all "Mekong subsistence peoples whose survival resources are being compromised in the name of profit and who do not possess the political freedom to publicly protest the decisions that are made over their river's future" (Mick O'Shea). Not least, the foundation seeks to make sure that global pressure is put on the Chinese authorities in particular who, with their plans for industrialization, are wantonly destroying one of the last more or less intact river landscapes in the world against the interests of politically voiceless subsistence people who depend on the river's natural health and vitality.

Photo
Dachaoshan Dam, the second large dam on the upper reaches of the Mekong, began operating in 2003. Michael O'Shea

Paper factories were among the larger industrial polluters of the environment until a few years ago. Paper production requires lots of water and lots of chemicals. But it doesn't have to be this way. Here's an example.

A farewell to wastewater–producing paper without polluting rivers

Some 300 million tons of paper are manufactured every year–and between 20 to 150 liters of water are needed to produce each kilogram, depending on the paper's type and quality. Most of this water gets polluted by the highly toxic chemicals used during pulping, bleaching, gluing, dyeing, and finishing processes.

Up until the 1950s, wastewater from paper factories, highly acidic and containing chlorine, was discharged more or less directly into rivers, lakes, and oceans. Thanks to more stringent environmental regulations, the paper industry today creates much less pollution. Water consumption has been significantly reduced, especially because water can be recycled several times within certain phases of production. But according to the World Bank, the wastewater from paper factories in highly developed countries still makes up 23 percent of all industrial wastewater (10 percent in underdeveloped countries), and this figure is rising. Thanks to new methods of production, some of the worst environmental toxins have been reduced or replaced by other less ecologically harmful chemicals–toxic chlorine has been replaced by hydrogen peroxide in the bleaching process, for instance.

The Julius Schulte Söhne paper and cardboard factory in Düsseldorf, working together with the Dutch facility contractor Paques Water Systems and Darmstadt Technical University, has developed a process that produces no wastewater whatsoever. Water used during paper production is treated and reused in a closed cycle. Firstly, dirty water is fermented in a special anaerobic tower reactor, a process which cleans it. Then it is decalcified in two ventilating reactors. The carbonate sludge formed here is returned to the production line and integrated into the paper.

According to experts from the Deutsche Bundesstiftung Umwelt, a German environmental foundation, this new production process doesn't exert any adverse influence on the quality of paper and doesn't require additional ecologically harmful chemicals or biocides for treatment. Last but not least–in spite of widely held biases that ecologically motivated improvements are to be had only at higher cost, this innovation pays for itself. The biogas derived

from wastewater fermentation is expected to cover the entire energy needs of the facility in the future. The company saves some 260,000 cubic meters of wastewater, and at the same time doesn't have to pay 400,000 euros in sewer fees each year. From an environmental point of view, the Rhine River is spared from being polluted by discharges of many tons of persistent harmful chemicals, which German and Dutch waterworks farther downstream would have to remove at great expense.

One might expect that such a uniquely advantageous system would take over the market for paper manufacturing technology in no time at all. But here legislation applies a brake. Practically everywhere in the world, policies on industrially polluted water still foster end-of-pipe wastewater treatment, and hence fail to provide a legal incentive for investing in manufacturing processes that can do completely without water or that are able to treat water for reuse, although

such technology is already available for numerous products.

A forward-looking innovation devised by a paper manufacturer in wealthy Germany could indeed be a vital blessing for populations in other countries. Poorer countries simply can't afford to treat river water which is as polluted as the Rhine in order to obtain drinking water. Countries in Southeast Asia and Latin America, highly dependent on rivers as reservoirs of drinking water, would be ideal locations for manufacturing processes that don't produce wastewater. But unless policies on water and legislative guidelines are redirected to keep industrial wastewater completely away from rivers in the future, the quality of drinking water, particularly in large cities, will continue to deteriorate.

Photo
In global terms, the pulp and paper industry is one of the biggest polluters. Durango, Spain.
Simon Fraser/Keystone

In many regions with little rainfall, especially in the Middle East, governments are tapping into deep-lying groundwater reserves to meet a rapidly growing demand for water. A risky venture—once these aquifers are used up, it can take thousands of years for them to be replenished. In the Sultanate of Oman, for instance, even banking on desalination plants will not suffice to permanently solve the problem of water shortage.

Water management tests the limits
—the case of Oman

Qaboos bin Said Al Said, the sultan of Oman, after violently taking over the realm and dethroning his father, has led his country from the Middle Ages to modern times in less than thirty years. Thanks to lavish state income from oil, the population has experienced rapid improvement in its standard of living. But better life quality means that demand for fresh water has also increased. The Sultanate of Oman, located on the Arabian peninsula, is an arid country with only 100 to 300 millimeters of rain per year, and fresh water is scarce. Rain and dew replenish groundwater, the country's only natural freshwater reserve, with an average of only 550 million cubic meters of water per year. The consumption of 645 million cubic meters of groundwater by 1995 had already surpassed this volume.

Groundwater reserves in the Sultanate of Oman are particularly threatened because of its unusual geological formations. Here, where about 94 million years ago the ocean floor slid over the Arabian plate to form the nearly 3000-meter high Jebel Akhdar

Mountains, underground water flows along abnormal paths. Underground reserves are usually connected, but in Oman they are concentrated in larger but isolated groundwater lenses. In the north of Muscat, the capital city, where demand for water is highest because the area is densely populated, these lenses are only six kilometers from the coast of the Indian Ocean and are threatened by the salination that occurs when seawater seeps in. If groundwater is pumped up too quickly, fresh water deep down gets mixed with salty seawater. Local overuse has already led to the salination of groundwater so that date palms near the coast are dying off.

To be able to use scarce water resources in a sustainable way, detailed knowledge is needed on the current use of groundwater and how it is replenished by rain and dew. Muscat, for example, draws its water from the Al Khwad Aquifer, a groundwater reserve that has three distinct horizontal layers. The presence of certain trace gases in groundwater enables scientists to determine how many years before and even where water has seeped into the ground. Al Khwad's top layer, 50 meters deep, contains "young" water that seeped down no more than fifty years ago and which can therefore be replenished in a few decades. The lowest layer, lying at a depth of about 300 meters, contains water that is between 15,000 to 24,000 years old, which means that water pumped up from this layer can be replenished only after thousands of years, making sustainable management practically impossible. The middle layer holds a mixture of ancient deep water and younger water from the top layer.

Using only the top layer of water can therefore be justified, but what degree of withdrawal is sustainable? To forecast the productivity of a groundwater reserve, scientists must know where this water originates. Hydrologists were able to localize the catchment area of Al Khwad based on information revealed by the relative content in water of oxygen isotopes ($^{18}O/^{16}O$) and hydrogen isotopes ($^{2}H/^{1}H$). Water in the top layer had not come from rains along the coastal plain, but had seeped into the ground high up in the Jebel Akhdar mountain chain. One option for increasing the available volume of groundwater would thus be to build small check dams to prevent surface water run-off in the mountains, thus improving on-site seepage.

Groundwater alone is not enough to supply the population of Oman in any case. Today a considerable quantity of water is already won from desalination facilities, surface waters, and wastewater recycling. But even this is not enough—per capita consumption of water in Oman is still rapidly increasing. At the same time, forecasts predict that the population will go up from 2.3 million (in 2003) to 4.7 million by 2025, and that prosperity will continue to increase. In 2000, inhabitants were already consuming 1,525 million cubic meters, and the expectation is that annual freshwater consumption will increase 63 percent by 2025.

But how the Sultanate of Oman will secure its water supply in the future is not just a question for scientists. The nation's government will have to fall back on all technology available if it wants to continue its present growth. It won't be able to simply make more and more water available. In arid regions in Australia, the United States, and Spain, we can see how the demand for water can be influenced (demand management)—if the political will is there (→ pp. 384). Luckily there is also a water resource that increases when consumption increases—wastewater. It seems inevitable that reusing purified wastewater in arid countries like Oman—even for drinking water, a rather unpopular use—will become a standard for water management in the future.

Photo
A hotel bathroom in Oman, taken in 2004.
Ian Berry/Magnum Photos

Fresh water can be found even in places where there aren't any springs bubbling or streams flowing. Canadian and Chilean scientists worked together on a unique project to develop a simple and inexpensive method for harvesting so much water from fog that it supplied an entire village of several hundred inhabitants with drinking water.

Water from fog

When scientists from Canada's meteorological services in 1980 collected samples of fog on Mount Sutton for chemical analysis, they never would have guessed that their work would help thousands of people, years later, to easily harvest good-quality drinking water. A visit by a delegation of Chilean scientists to Canada gave rise to a cooperative effort lasting seventeen years which led to the construction of the world's first major fog collection device in the Chilean fishing village of Chungungo on the extremely arid Pacific coast (60 millimeters of rain per year). Mist netting was installed in cliffs high above the village to harvest fog rising from the ocean. Over the years, Chungungo became the synonym for fog collection.

Fog is defined as a collection of very fine droplets of water suspended in air close to the ground. As soon as these droplets are blown against a solid surface, they precipitate and form larger drops of water. Fog collectors make use of this property. Made of rectangular nylon or polypropylene netting, they are suspended from vertical poles about 1.5 meters above the ground at a right angle to the direction of prevailing winds. Droplets of fog precipitate on the mesh and are collected in a cistern at the bottom of the netting.

By the early 1990s, this technology had matured to the point that the Canadian government funded the construction of a drinking-water pipe 7 kilometers long from the mountain ridge down to the coastal village of Chungungo. More than one hundred households now had so much running water from fog that they could even cultivate fruits and vegetables. Some one hundred fog collectors installed on the mountain ridge provided an average of 15,000 liters per day, and on good (foggy) days, up to 100,000 liters. The system functioned for about ten years and this reliable supply of water enabled the village to expand from 300 to 600 inhabitants. Several thousand people live in Chungungo during the summer and it boasts gardens with fruits and vegetables. The income from this produce brought electricity and a gasoline pump to the village.

But the fog collectors gradually fell into disrepair and by 2003 none of them worked anymore. Faced with a choice between replacing the collectors or building a different water-supply system, the local population opted for the latter. There are plans to install a seawater desalination plant or to pipe water in over a long distance, both options requiring investments of at least 1 million dollars. Water presently has to be brought in by tanker truck from far away at great expense to the village, which is just the way things were before the fog collectors were installed in May 1992.

Fog collectors continuously and reliably supplied water to Chungungo in Chile's coastal desert for ten years, enabling the village's growth and development in the first place. Today, with more money around thanks to the collectors, inhabitants are abandoning this system. Time will tell whether it was the right decision to turn away from fog, which reliably supplied drinking water locally by using simple technology and without consuming energy.

Thousands of villages around the world subsist in a similar desert-like environment with regular wind-blown fogs. In such locations with low rainfall, a single square meter of netting suspended in the path of wind-blown fog will collect some ten liters of water per day for drinking and irrigation. Fog harvesting is an excellent option in the absence of alternative sources of water and wherever technology such as desalination is too expensive. It is currently being used in twenty-two countries around the world.

Info
www.fogquest.org

North American rivers have been extensively impounded in the past century. Many of these dam structures are now approaching the end of their life spans, either because their reservoirs are filled with sediment or the aging dam infrastructure becomes unsafe. Some dams have already been removed, yet dealing with the immense masses of accumulated sediment is a considerable technical and scientific challenge.

When dams get old: Dam Removal in Western North America

In the United States today, there are approximately two million dams and weirs of varying sizes, with over 80,000 dams above 1.5 meters high (Graf 2005). These structures range in size from small diversions for irrigation or hydroelectric generation to enormous reservoirs impounding many times the river's average annual flow. Most dams were designed with life spans of fifty to a hundred years, and as the dam infrastructure ages and reservoirs fill with sediment, there are increasing calls for dam removal. In general, it's much easier to remove small dams than large ones, because the sediment trapped behind large dams can pose problems if it washes downstream, and because it's more expensive to compensate those who benefit from the dam's functions.

Many rivers in western North America support runs of anadromous salmonids, i.e. salmon and trout that migrate upriver from the sea to spawn, and back to the sea as young fish (juveniles). These species depend on a continuous, undammed river for their cycle of reproduction, and dam construction has resulted in anadromous fish disappearing from many rivers in western North America. Consequently, most dam removals in this region have been motivated by restoring fish populations or by safety concerns because of aging structures.

In understanding the effect of dams on river systems and the issues raised by their removal, scale is key. All dams interrupt the flow continuum of rivers (reduced longitudinal connectivity), impeding migration of fish and other aquatic organisms. To alleviate that effect, smaller dams can often be fitted with fish ladders to permit the upstream migration of adult salmon and trout. But even if fish ladders work for adults, young fish may not be able to successfully migrate downstream because reservoirs lack strong downstream currents that provide cues for the juveniles, or because the juveniles are injured or killed as they pass through turbines or release structures.

Most dams removed to date have been small structures which could be removed with minimal or transient effects as they did not trap large volumes of sediment, did not substantially alter flood flows, and whose economic importance had declined or for which substitutes could be found. Removing these structures usually restores unimpeded fish migration to the river ecosystem.

By contrast, large dams are more likely to substantially alter a river's flow regime. In addition to restoring free fish migration (longitudinal connectivity), their removal would significantly change the river's flow patterns, restoring flow volume and its seasonal variations to the natural regime important for the ecological health of rivers (Poff et al.1997). However, the reservoirs of large dams are efficient traps for the particulate matter transported by rivers. Sediment accumulated over decades thus sits at the bottom of these reservoirs. When such dams are removed, the decisive question is what happens to the sediment. Also larger dams typically support more economic activity, so their removal raises the rather political issue of finding alternatives and compensating affected parties.

In California alone, there are over 1,400 dams higher than 7.6 meters, mostly built in the early to mid 20th century. California has the highest number of dam removals to date, with over seventy dams removed, mostly small structures and most to improve fish passage. Of the thirty-seven projects with available size data, the average height of the dams removed in California was 5 meters, ranging from 1.5 to 17 meters (Gilbreath 2006). The case studies below include two from California and one from Washington State, illustrating issues facing managers as they move forward with dam removals.

SMALL
Small dams, though impounding only a small fraction of the annual runoff, can substantially affect base flows—in some cases

temporarily drying out the rivers by diverting the entire base flow. On the other hand, due to their low water-storage capacity, small dams have little or no effect on the river's flood flows. These dams trap bedload sediments (gravel and sand), but the volumes trapped are commonly small enough that they can be mechanically removed, or if the reservoir sediments are left in place, they are normally transported downstream by natural flows in less than a decade after dam removal.

An example of a recent small-dam removal was on Butte Creek (380 km2), one of the few remaining streams in the Sacramento River system (California) still supporting healthy runs of spring-run Chinook salmon (Oncorhychus tshawytscha). Numerous small agricultural diversion dams built in the early 20th century created partial barriers to upstream migration by adult salmon. From 1993 to 1998, five dams ranging in height from 2 to 5 meters were removed, and five other dams were retrofitted with fish ladders so they no longer impeded fish migration. The restoration program also included both the installation of fish screens to prevent salmon from being swept from the stream into diversion canals and the acquisition of land along the creek, at a total coast of approximately $35 million. Fish counts in subsequent years showed substantially increased numbers of salmon in upstream reaches, increasing from a few hundred adults annually in most years prior to 1993, to over 7,000 in six of twelve years from 1994 to 2005, and making Butte Creek one of the most prolific spring-run salmon streams in California (Friends of the River 1999). Notably, the dams on Butte Creek were removed in a way that still allowed the diversions to occur, i.e. there was no impact on agricultural water withdrawals and no need for compensations. Thus the restoration program improved conditions for migrating salmon without conflicting with water use established by agricultural interests–considered a "win-win" situation.

MEDIUM
In the geologically unstable California Coast Ranges, much material is eroded and flushed away by rivers, and several small water-supply reservoirs have filled with sediment and hence lost their storage capacity and usefulness. Matilija Dam on Matilija Creek, Ventura River (142 km^2 catchment) is a concrete dam 50 meters high completed in 1949 to impound water for irrigation. It was constructed with poor-quality concrete and as a result is decaying

and unsafe. Approximately 4.5 million cubic meters of sediment have accumulated behind the dam. The dam blocks access to what were historically some of the best spawning and rearing habitats for anadromous steelhead trout (O. mykiss), one of the southernmost native runs of this species. Removal of the dam is thus expected to open up important fish habitats and eliminate a safety hazard. The biggest issue has been how to handle the accumulated sediment. Simply removing the dam and allowing accumulated sediments to be transported and deposited downstream may raise the bed, reduce flow capacity, and increase flooding. Potential liability for damage to property along the lower reaches of the river might be expensive, especially in light of the astronomically increasing property values in the region.

The project scope includes removing (by slurry pipeline) approximately 1.5 million cubic meters (a third of total deposits) of fine sediments from behind Matilija Dam approximately 8 km downstream to slurry disposal sites. The remaining 3 million cubic meters of deposited sediments are to be left in place, but reshaped with heavy equipment into terraces. A 30-meter wide meandering fish passage channel through the former sediment deposit will be constructed. Modeling studies suggest the renewed inflow of upstream sediment to the river (after decades of artificial sediment-trapping by the reservoir) will reverse the erosion trends caused by the dam in the downstream reaches of the Ventura River and may restore a more natural equilibrium of deposition and erosion of sediments in approximately a decade. The questions of how to manage the accumulated sediments and what the impact will be of releasing them to downstream reaches, in which the channel is flanked by expensive property, are of utmost economic and political importance as removals are considered for other dams in the California Coast Ranges, such as San Clemente Dam on the Carmel River (324 km^2) and Searsville Dam on San Francisquito Creek (38 km^2).

LARGE
The Elwha River (draining 700 km^2 of the Olympic Mountains of Washington) has two dams, the 64-meter high Glines Canyon Dam impounding a 50-million-cubic-meter reservoir, and the Elwha Dam impounding a 10-million-cubic-meter reservoir. These dams are operated as run-of-the-river structures to generate hydroelectric power, so they have relatively little effect on the river's seasonal flow patterns, but they have

trapped about 13 million cubic meters of sediment. As a result of sediment being trapped by the reservoirs instead of being deposited along the river and its estuary ("sediment starvation"), the riverbed downstream has coarsened over the decades and the coastal zone at the river mouth has experienced accelerated erosion (Gregory et al. 2002).

Prior to dam construction in 1910, the Elwha River supported runs of all five species of Pacific salmon native to North America, steelhead trout, and dolly varden (Salvelinus malma). The potential to re-establish such a rich and diverse fish community, and the willingness of the dam owner to sell the dams and substitute other sources of power, have resulted in making this the largest dam removal pending in North America. The US Congress passed legislation to remove the dams in 1992, and the federal government acquired them in 2000.

Unlike the Matilija and other California dams discussed above, the reach downstream of the dams on Elwha River is flanked mostly by rural lands. Thus, even though the volumes of sediment stored are much larger than in most other dams considered or slated for removal, allowing some stored sediment to move downstream would be less likely to affect infrastructure than in more heavily settled areas.

Modeling demonstrated that the deposited sediment would not all mobilize at once, but would gradually erode. There is wide support for the removal of the Elwha dams because of the potential to increase anadromous salmon populations and to restore sediment supply to the lower river and coast.

LEARNING FROM DAM REMOVALS

Given the likely increase in dam removals in the future, the removals undertaken in the past and today can been seen as "pilot" projects from which we can learn lessons to improve future planning and implementation. Unfortunately, most dam removals to date have been poorly documented. In an effort to compile data from dam removals in an objective, easily accessible archive, the University of California Water Resources Center Archives has created the Dam Removal Data Base (http://www.lib.ber keley.edu/WRCA/damremoval/index.html). As dam removals are completed, summary data are entered into the database and relevant documents added as downloadable pdf files. Every dam removal has its own unique set of physical, ecological, and institutional characteristics, but with the improved documentation of pre-project conditions, clear statement of goals, and post-project monitoring, it should be possible to learn from ongoing dam removals to improve practice in future removals elsewhere. Small-dam removals (less than 5 meters high) are most common, and offer the greatest opportunity to study dam removal due to the large number of cases. Removing large dams will usually require dealing with millions of cubic meters of accumulated sediment, a task vastly more challenging and costly, and requiring detailed, site-specific studies.

References
– Friends of the River. 1999. Rivers Reborn: Removing Dams and Restoring Rivers in California. Available at http://www.friendsoftheriver.org/Publications/PDF/RiversReborn.pdf.
– Gilbreath, Alicia. 2006. Dam removal in California: lessons learned through the post-project appraisal. Master's thesis, Department of Landscape Architecture and Environmental Planning, University of California, Berkeley.
– Graf, W.L. 2005. Geomorphology and American dams: The scientific, social, and economic context. Geomorphology 71, p. 3-26.
– Gregory, S., H. Li, and J. Li. 2002. The conceptual basis for ecological response to dam removal. Bioscience 52:713-723.
– Heinz Center for Science, Economics and the Environment (Heinz Center). 2002. Dam Removal: Science and Decision Making. Washington DC: Heinz Center. Available at http://www.heinzctr.org/NEW_WEB/PDF/Dam_removal_full_report.pdf.
– Kondolf, G.M., A. Boulton, S. O'Daniel, G. Poole, F. Rahel, E. Stanley, E. Wohl, A. Bang, J. Carlstrom, C. Cristoni, H. Huber, S. Koljonen, P. Louhi, and K. Nakamura. Process-based ecological river restoration: Visualising three-dimensional connectivity and dynamic vectors to recover lost linkages. Ecology and Society. (in press)
– Poff NL, Allan JD, and Bain MB. 1997. The natural flow regime: a paradigm for river conservation and restoration. BioScience 47: 769-84.
– U.S. Army Corps of Engineers, Los Angeles District and Ventura County Watershed Protection District. 2005. Matilija Dam Ecosystem Restoration Project, Project Management Plan. Available at http://www.matilijadam.org/pmpfinal.pdf.

Photo
Due for dismantling: The Elwha Dam in the state of Washington, USA. Photo taken in 1995. Matt Kondolf

---

Just a few decades back, several million children died of diarrhea in the Third World every year. It was only thirty years ago that modern science discovered an effective remedy—a simple solution of sugar and salt in water. But this finding is hardly new. Sushruta, an ancient scholar of India, recommended this recipe to patients 2,500 years ago.

The salt of life—oral rehydration therapy

As recently as twenty years ago, acute diarrhea was the most common cause of death among children. In the 1980s, about 4.6 million children died each year from the loss of fluids caused by diarrheic illnesses. In an advanced stage of illness, even frequent drinking will not make up for fluid loss because body tissue is no longer able to absorb fluids rushing through the digestive tract. In those days, it seemed that the only option was to bypass the digestive system by infusing fluid directly into the blood system—a traumatic and painful method that can hardly be put into practice with sick children. Added to that, intravenous infusion can be done only by medically trained personnel—which most affected patients don't have access to.

Scientists in Bangladesh and India in 1968 discovered that adding salt and sugar to water in certain proportions enabled water to be absorbed by intestinal walls. Simple and unspectacular as this method is, it must nevertheless be regarded as the biggest advance in medical research in the 20th century, measured by the number of lives it has saved. It was finally possible to quickly and easily help patients suffering from dehydration caused by diarrhea—with one teaspoon of salt and eight teaspoons of sugar dissolved in a liter of water. This simple recipe has since then saved many millions of people from certain death.

During Bangladesh's war of independence (1971), cholera broke out severely in refugee camps. Health workers decided to use the oral rehydration method on a large scale for the first time. Of the 3,700 patients who were treated, 3,552 survived.

Since 1979, oral rehydration for treating diarrheic illnesses has been an essential component of the UNICEF program, the United Nations' aid organization for chil-

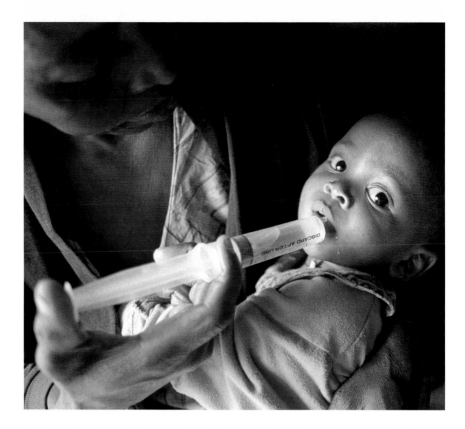

dren. Today the number of children under the age of five who die from the consequences of diarrhea is about 70 percent lower than before. Hundreds of millions of ready-to-use packets with salt and sugar are prepared each year. They cost about 10 cents apiece–10 cents that can prevent a child from dying of diarrhea and dehydration in more than 90 percent of cases.

Due not least to UNICEF's strategy, no large epidemics of diarrhea broke out in the weeks following the tsunami disaster in December 2004. UNICEF trained volunteers in the most heavily affected areas to draw attention in towns and villages to the most important preventive measures. UNICEF representatives using minibuses and loudspeakers informed people at rallies about oral rehydration.

It must seem incredible to us that a prescription of one teaspoon of salt and eight teaspoons of sugar dissolved in one liter of boiled water could really be a 20th century discovery. Indeed, as long ago as 500 BC, an Indian scholar and father of Ayurveda, Sushruta, prescribed to his cholera patients suffering from diarrhea, "lots of lukewarm water in which rock salt and molasses have been dissolved, or purified water combined with rice-meal gruel" (Sushruta Samhita III, Verse II). Perhaps Sushruta's simple formula was too banal for 20th-century medical researchers to take seriously.

Photo
Rehydration of a child suffering from cholera in KwaZulu-Natal. Photo taken in 2000.
Cobus Bodenstein/Keystone/AP

Sometimes it's really easy to solve a problem. Water contaminated with germs can be easily disinfected if it is exposed to the sun for several hours in a plastic bottle. Several aid organizations and foundations are making a big effort, especially in Latin America, to popularize this procedure, which was originally discovered by Swiss scientists.

Solar disinfection of water
−drinking water in six hours

The lack of access to safe drinking water and sanitation facilities still endangers the health of many millions of people. Every day about 6,000 people around the world, mostly children under the age of five, die of diarrhea. The World Health Organization reckons that 1.1 billion people must survive without access to safe drinking water.

Following some simple rules of hygiene—washing hands after defecating, regulating the disposal of excreta, and safely storing drinking water—does drastically reduce the number of cases of illness. But these measures are of secondary importance if available drinking water is not of good quality and contains pathogens. In many places, even faucets in a house and public taps can deliver water that is questionable because sources are polluted, pipes have been poorly maintained, or mistakes have been made during the withdrawing, transporting, or storing of drinking water.

It is rarely practicable to boil water–wherever people don't have access to safe water, they usually don't have firewood either, or money for fuels such as kerosene or gas. Chlorinating drinking water to disinfect it can be ruled out because the chemicals needed are often not available in remote regions. Moreover, most people reject chlorinated drinking water because of its unpleasant taste.

The idea of solar disinfection (SODIS) does however raise hope for a basic change in disinfecting drinking water. It is not dependent on expensive chemicals and fuels–it takes advantage of the fact that solar heat and ultraviolet (UVA) radiation work together to effectively kill pathogens. The recipe is extremely easy–fill a clear plastic bottle with water to be purified and lay it out in blazing sunlight for six hours. Bacteriological tests show that water is hygienically pure after this time. Is solar disinfection the long-sought solution for purifying water at home?

The application of SODIS has progressed farthest in Latin America, where 60 million people don't have access to safe drinking water. This simple way to purify water is promoted and supported by the SODIS Foundation. The foundation has a widely branching network of partner institutions. When cholera epidemics broke out in a number of South American countries in the 1990s and claimed the lives of several hundred people, it became imperative to have a simple method for purifying drinking water. Since then, SODIS has been tested and implemented in seven Latin American countries, taking various sociocultural aspects into consideration. The work with local people has been carried out mostly by NGOs, but more and more health authorities also encourage the application of SODIS. Experts in Latin America estimate that by the end of 2005, SODIS was being regularly used by some 300,000 people.

A school project initiated by Project Concern International and the SODIS Founda-

tion in seven communities in the highlands of Bolivia is said to be exemplary. Bolivia's Altiplano lies at an altitude of 3,200 to 4,200 meters above sea level where the climate is cool and solar radiation is intense. The population, for the most part very poor, lives according to regional and traditional customs and some communities (Larampujo, Alcamarca, Tarucamarca, Realenga, Sora Sora, Chiwirapi, and Janco Janco) are very difficult to reach. Because schools are the ideal platform for spreading information under these conditions, SODIS became a permanent part of the teaching program. Regular visits to homes and direct advice ensured that the method was correctly applied. Today, 85 percent of the local population (2,210 persons) report using SODIS every day. Some women have even knitted bottle holders from colored wool so they can carry treated water to work in the fields.

A second application area for SODIS is in disaster aftercare. When an earthquake affected the Arequipa region in Peru in 2001, houses were destroyed along with a substantial section of water piping. Some 416 families who lost their safe water supply as a result were given instructions on daily hygiene and SODIS with the help of the local Red Cross and with funding from Fondo de las America. Health brigades elected by the population helped these families develop new habits in their daily lives. This participatory approach contributed greatly to motivating the population, and by the end of the project, 75 percent of those affected reported using SODIS daily.

To date, more than two million people in 20 countries around the world have become familiar with SODIS and at the same time become familiar with improved practices in hygiene. Health surveys show that diarrhea has dropped by an average of roughly 50 percent among SODIS users. In recognition of this service, those responsible for the SODIS project received in October 2004 the Energy Globe Award, a prestigious environmental prize.

Infos:
www.sodis.ch

Photo
Sunlight is used in Bolivia to sterilize drinking water in PET bottles. Keystone/Helvetas

This is the paradox of conventional wastewater management—drinking water is used to flush human excreta into the sewer system where it is further diluted by household wastewater from kitchens and baths. As a result, wastewater treatment plants must employ ever more complex process engineering to remove pollutants and recycle valuable nutrients. If we had to redesign wastewater disposal systems based on what we know today, we would certainly not opt for the same package of technologies.

Novaquatis—a building block for innovative sanitation technology

Today's prevailing wastewater management is not based on an integral design but is rather the result of a long technological development that began with the construction of sewers in the 19th century. Sewers were originally built for carrying away rain and wastewater. Drains from flushing toilets were later connected to sewer systems, which at that time still transported wastewater directly to nearby rivers and lakes. The resulting massive pollution of these bodies of water gave rise to the construction of wastewater treatment plants, first to remove visible particles and later to remove biodegradable organic material. During the 1970s, it became clear that not only organic matter, but also nutrients in wastewater were causing grave problems. Lakes became overfertilized with phosphorus and some reached the stage of biological collapse; the ammonia discharged into rivers, for instance, proved to be toxic for fish. Consequently, treatment plants were improved to remove phosphorus and convert toxic ammonia into nitrate.

More recently, nitrate itself was recognized as a cause of overfertilization in marine coastal areas. Treatment plants were upgraded again to enable denitrification, a process now required by law in many areas of Europe to protect coastal waters, especially the North Sea. However, there seems to be no end to the surprises found in wastewater. Scientists have recently discovered that residues from pharmaceuticals, hormones, and other biologically active substances discharged into municipal wastewater are present in rivers. Dealing with these substances may call for making yet another improvement in wastewater treatment plants.

After more than a century of corrective upgrading, wastewater infrastructures in industrialized countries have become increasingly complex and costly. Some 80 percent of the costs of total wastewater disposal are due to the fact that sewer systems require continuous maintenance and renewal.

Developing and emerging countries today face the question of whether they should adopt this far-from-perfect system. Constructing a sewer system for the first time is a complex and highly expensive venture. Making large investments of this kind is often nearly impossible for municipalities, especially in rapidly developing cities that face chronic shortages in capital. Additionally, removing excreta by means of a sewer system requires using large quantities of water that are simply not available in countries where water is scarce. But are there alternatives to the western system of wastewater disposal? Yes, there are. Several alternative approaches have been researched and tested, although most of them are still in the pilot phase.

One idea with excellent potential is urine source separation, which keeps urine separate from feces in special NoMix toilets, using little or no water and collecting urine almost undiluted. Urine contains the largest share of the nutrients excreted by humans. If urine were kept entirely separate from wastewater, treatment plants could do without today's expensive and complex process of nutrient removal and return to the simpler and cheaper wastewater treatment of the 1950s. Since many of the pharmaceuticals consumed by humans are also excreted in urine, these substances, which are potentially harmful ecologically, could also be kept out of wastewater and away from receiving waters. The nutrients in urine (nitrogen, phosphorus, and potassium) could be used as fertilizers in agriculture. Moreover, the NoMix process is of value in countries with little water because very little or no water is needed for flushing urine. It is also much easier to treat wastewater to be reused for other purposes if it does not contain salty urine to begin with. Urine source separation may be a first step in an entirely new concept for managing

wastewater. However, it should be noted that wherever the infrastructure already boasts a fully developed network of sewers and efficient treatment plants, urine-separation technology is practicable only when it can be integrated successively into the existing system.

But the situation is different wherever a wastewater infrastructure is still being built. In the region around Kunming in China, for instance, many localities do not have a sewer network. Urine source separation in dry toilets is one of the options being considered there by local authorities. The larger metropolitan area of Kunming, currently boasting a population of 2.6 million inhabitants forecast to reach 4.5 million by 2020, lies on the shores of a shallow lake, Lake Dianchi, which is already heavily overfertilized. The construction of several wastewater treatment plants between 1988 and 2001 did not reduce the nutrient pollution of the lake. Untreated wastewater, and run-off from agriculture and phosphorus mines, continue to exacerbate overfertilization.

An ancient Chinese tradition of sustainable nutrient management, including urine collection for agriculture, may provide a good basis for reintroducing old practices in a new technological guise, in this case as urine-separating dry toilets. The world's largest pilot project was set up in Guangxi province with the installation of 150,000

urine-separating dry toilets. These toilets work entirely without flushing water–something unusual for western visitors. Separately collected urine is spread directly on fields as fertilizer, and feces collected as dry material are mixed with ashes, stored for hygienization and later also used in agriculture. In this way, hardly any nutrients or pathogens reach wastewater. A pilot project in a village close to Kunming, with one hundred dry toilets, was well accepted by the local population. Swiss scientists, as well as engineers and social scientists from Kunming, local decision-makers, and the inhabitants concerned all participated in the project. Impressed by its success, the local administration is considering installing 100,000 urine-separating dry toilets in Kunming in coming years.

The question is whether installing a system with urine-separating dry toilets in downtown Kunming will be accepted by a population now oriented towards western standards. A new sanitation system can make its entrance in an urban setting only if it satisfies the highest demands in comfort. A water-flushing NoMix toilet being used today in smaller pilot projects in Europe seems more likely to be accepted by the urban population.

NoMix toilets available today are still not fully developed technologically, and leading manufacturers of sanitation equipment hesitate to invest in the further development of such toilets for the future. Western markets will remain small as long as existing technology, in spite of its drawbacks, is seen as the best solution. This can change if the benefits of NoMix technology are recognized in regions now going through the process of designing infrastructures for wastewater management. Although urine source separation still calls for further technological development and needs testing in pilot projects, it may be cheaper and more efficient in the long run to follow this innovative path instead of relying on conventional western end-of-pipe technology.

Based on a text by Judit Lienert
and Tove A. Larsen

information
www.novaquatis.eawag.ch, novaquatis@eawag.ch

Photo
Urine is collected separately, almost entirely undiluted with flush water, in nomix-toilets. Roediger

Countries in the developing world cannot afford to build wastewater management systems like those normally found in northern industrial countries. In Kumasi in Ghana, the United Nations Development Program, together with the World Bank and the Ghanian government, financed a successful pilot project intended to meet the special needs of poor cities.

Solving sanitation problems in Kumasi, Ghana

One of the most important and difficult tasks confronting urban authorities around the world is the management of excreta, wastewater, and garbage disposal. Authorities in developing countries are often unable to shoulder these tasks. Financial resources, a sustainable tariff system, the political will, and enforceable regulations are frequently missing, along with the needed skills and organizational know-how.

Conventional solutions for planning and developing urban waste and wastewater management based on the model of European cities led to ambitious projects for building sewer systems and sewage-treatment plants that ultimately weren't sustainable because of the special problems prevalent in Third World countries. Aside from the financial burden they posed, these projects had serious health consequences in many cities, especially for poorer populations.

This dilemma gave rise to the World Bank and other international development organizations creating a planning tool called Strategic Sanitation Planning (SSP). This tool was tested in a project attempting to improve the sanitation and hygiene situation in Kumasi, Ghana. A multi-component pilot project was devised to sustainably tackle the challenges posed by the need for improving the management of excreta, wastewater, and solid waste. The project was co-financed by the United Nations Development Program, the World Bank, and the government of Ghana.

Kumasi, a middle-sized city in Ghana with a population of 700,000 at the time the project began, was selected for the project because the majority of its inhabitants didn't have access to appropriate, private sanitation facilities. The goal was not to directly subsidize households in Kumasi but to develop model solutions for cities in which people had become ill as a consequence of poor sanitation, most notably because of the inadequate disposal of excreta.

There were no hygiene facilities in three fourths of Kumasi's households when the

project began. Forty percent of the population used public latrines and 5 percent of the population relieved themselves out in the open. In 25 percent of households, bucket latrines were used that had to be emptied by hand every day. Five percent used simple pit latrines and 25 percent flushing toilets linked to septic tanks. The disposal of sludge collected from latrines and septic tanks by suction vehicles was also inadequate. More than 90 percent of the collected excreta ended up in nearby rivers or on unused plots of land. Water pollution and the risks to health were intolerable.

Kumasi's city government for years had tried different strategies to solve its sanitation problem, without lasting success. The pilot project, to be planned and seen through in line with SSP principles, presented a unique opportunity for improving the situation on a long-term basis. A strategic plan was designed with technical and organizationally appropriate solutions for managing human and solid waste. The plan took into consideration the differing conditions and needs of each of the city's settlement zones, and was carried out with the support of UNDP and the World Bank. Households' preferences and willingness to pay were taken into account. The time frame for the project was limited to a maximum of 10 to 15 years, and the overall project was broken down into many sub-projects that could be tackled independently. Priority was given to

test projects and those parts of projects that could be put into action right away.

Disposal technologies were adapted to relevant living conditions since today, as in the future, only a small part of the city can be served by sewer systems, the standard technology in industrialized countries. Excreta disposal and management options included:
– ventilated and easy-to-clean latrines for households with low income;
– flushing toilets and septic tanks for new government buildings and houses belonging to the upper classes;
– sewer systems with simplified construction standards only for those apartment buildings in a few densely-built areas where latrine construction was not technically feasible, and where space was lacking for effluent disposal and did not allow access to suction vehicles.
Wherever on-site sanitation options (latrines or septic tanks) were seen as the solution, inhabitants could choose between different types, depending on their income and their cultural habits.

Kumasi became a model project. The sanitation and hygiene situation in the city improved within a few years and the project was extended. Achievements went far beyond the original goal. Authorities in other cities in Ghana and in surrounding countries (Ouagadougou, Burkina Faso; Conakry, Guinea; Cotonou, Benin) benefited from the lessons learned during Kumasi's planning of urban sanitation using SSP guidelines.

This new approach to planning and building sanitary facilities, including the disposal of excreta and wastewater, and focusing on the participation of the inhabitants concerned, has proven successful. A substantial, additional benefit gained from the Kumasi sanitation project was the training of professionals who are today in a position to look after complex urban sanitation tasks. At the same time, institutions were put into place or restructured in a way that allowed the city to safeguard public health and protect the urban environment on a sustainable basis under the very challenging conditions prevailing in a developing country.

Photo I
Faecal sludge from latrines and septic tanks is treated separately in a special sewage treatment plant. Kumasi, Ghana. Eawag/Sandec

Photo II
A poor district in Kumasi, where latrines are currently the best way of disposing of faeces. Eawag/Sandec

Goulburn, a city of 25,000 inhabitants in Australia's interior, is in danger of literally going dry in face of a drought that has persisted for more than two years. This desperate situation has led the district council to plan a project that could pave the way for dealing with water scarcity in other parts of Australia as well.

Hope for Goulburn

"My neighbor has been taking a shower for 20 minutes, I can hear it! This is awful! Do something about it!" says an agitated voice coming from the phone held by an employee at Goulburn's city administration. It's hardly a surprise that the person at the other end of the line is getting excited. It really isn't appropriate to stand under the shower for a long time these days in Goulburn. The level of water in the Pejar Reservoir, the larger of two dammed reservoirs that supply drinking water to the city, is at less than 6 percent of full capacity. Pictures of dried-out earth that should be lying deep under water are seen regularly in the Australian media. Usable water in both lakes together adds up to barely 12 percent of total capacity. People are talking about the worst drought in a hundred years.

WELLS AND EMERGENCY PLANS
Drastic limitations on water consumption have been in force for months. Private households and industries are allowed to use only what's barely necessary. Some recent rains have pushed back by about eight weeks the moment all are dreading. If things don't fundamentally change, the city could go completely dry in the foreseeable future. If drastic measures to save water hadn't already been enforced it would have gone dry in November 2004.

But it won't come to that. The administration in the shire of Goulburn-Mulwaree has been looking at emergency plans. The most extreme version would be to transport water to the city in trains. One train per day could provide the population with a minimum for survival and keeping up the most essential public services, but there would be nothing left over for the business sector–with all ensuing implications. Matt O'Rourke, the district administration's expert on water, would prefer not to think about the costs and logistical problems of distribution that would be involved.

Right now hopes are resting on tapping into a groundwater reserve located near the city. But O'Rourke believes that knowledge of the groundwater-system hydrology is insufficient to allow making a reliable prediction on how much time the city would effectively gain from mining this resource. It might be able to get by for several months, maybe one year, but it would have to find a sustainable solution by the end of that time. The city could at least hope for average rainfall to return (650 millimeters per year in Goulburn), but it couldn't take that into account in its planning anymore.

IS IT "ONLY" DROUGHT OR HAS
THE CLIMATE BEGUN TO CHANGE?
O'Rourke doesn't know whether the drought that has been choking broad parts of Australia's Southeast for more than two years is only a temporary phenomenon or a foretaste of what could happen if the climate is indeed changing, not just for Goulburn but for a large part of Australia.

People in Goulburn already know very well what it means to use water sparingly. The strictest possible restrictions short of an emergency have been in force since October 2004. The city's daily consumption of water has dropped from 13 million to 5 million liters since water-saving measures were installed. About 120 liters per person per day are presently being consumed in private households. Industries, namely the slaughterhouse and the wool-washing factory, both high users of water, have lowered their consumption by 30 percent. A positive side effect for the wool-washing factory at least is that it has been able to patent its water-saving technology. If Australia does indeed dry out, others will benefit from its know-how.

Every other use of fresh water outside of homes in Goulburn is banned. Washing the car and watering the garden are memories of a long-lost past. People with swimming pools can observe levels dropping as water evaporates. In many sports clubs, grass playing areas have dried out and hardened, making them unusable because the danger of suffering injuries is too high. The city's outdoor public pool was in use only during the three hottest summer months–then its water was pumped into the indoor pool. In

the meantime, even the indoor pool has closed because so many chemicals have accumulated in the water that it has to wait for better times when it can be diluted once again and treated to meet its original purpose. The water has been pumped back out to the outdoor pool.

Paul Stephenson, Goulburn's mayor, says that the population continues to use water-saving measures and has been very cooperative. But the administration must come up with results soon because the civic mood could turn. The first cracks are already visible. Because the wool-washing factory and the slaughterhouse, two important employers, have been forced to reduce water consumption, production has gone down. This in turn has had a direct effect on

the level of employment and indirectly on the general mood in the city. There are fewer and fewer leisure activities because the most popular sports clubs have cancelled their activities or moved away, at considerable cost, to the Australian capital Canberra, an hour's drive away.

Drinkers of beer everywhere in town are served their favorite beverage in plastic tumblers. Although this is an outrage for any real Australian, people go along with it because the bartender doesn't have to waste water unnecessarily by using a dishwasher.

THE NEED TO RETHINK
But not just the beer-drinking public has to swallow what used to be unthinkable–the

population at large has to as well. Realistically, the only options left are to recycle wastewater or to just not have enough. The district council is therefore planning to collect a substantial percentage of wastewater, clean it to "gray water" and pump it back to the catchment area of the reservoirs. There a system of ponds will be laid out in suitable terrain and serve as a natural filter plant. A similar system, which is used however only to clean water for industrial purposes, has already been installed in the Adelaide metropolitan area, which has more than a million inhabitants. If Goulburn does opt for treating wastewater to reach drinking-water standards, it will be a pioneer in Australia, not only for technical reasons but, even more so, in an ideological sense.

"Purified water–no thanks!" is the response in surveys, again and again. Fresh water has become a rare commodity in many parts of Australia, but no one wants to give up old habits. Abundant and cheap water, and abundant and cheap energy, seem to be taken for granted by many city dwellers. The message that reality could soon change on the driest inhabited continent on Earth,

where pro capita emissions of greenhouse gases are among the highest in the industrialized world, has apparently not yet got through everywhere.

The government of New South Wales, the most heavily populated state in Australia, recently announced that a seawater desalination plant would be built for the Sydney metropolitan area, which boasts a population of four million people and is also plagued by water scarcity. This decision is highly controversial. Critics see a need to explore more thoroughly all the options for treating and purifying water to be used in areas where drinking water is being unnecessarily squandered today. Sadly, the rethinking needed to deal with an ever more costly resource has not even gotten off the ground.

Translation of a text
by Rudolf Hermann

Photo
The reservoir (almost empty) of drinking water at Pejar Dam in June 2005. It is the main source of water for Goulburn. Australia. Alan Porritt/ Keystone/EPA

In an unparalleled economic experiment in 1989, Margaret Thatcher's government issued a decree to privatize the entire waterworks and wastewater facilities of England and Wales. This radical measure was based on the idea that the private sector works more efficiently, that private enterprise is better able to raise the capital needed for investment in infrastructure, and that privatization would encourage competitiveness. Seventeen years later, what is the situation now?

The privatization of water utilities in England–not a success story

The privatization scheme in England and Wales–Scotland and North Ireland were unaffected–was regulated centrally by the government, which not only handed over management to private operators, as had been common practice in France for decades, but also sold all facilities, mains, and real estate belonging to water and wastewater utility companies. Ownership of all assets belonging to the water-distribution and wastewater-disposal systems was transferred to private companies, which were also granted concessions for an initial period of twenty-five years.

The project was a matter of prestige for the Thatcher government. To make the privatization of water supply a successful venture, the state spent more than 8 billion pounds (13.2 billion euros) of tax money to pay off the debts of the water companies about to be privatized. It also made available to them an additional 2.6 billion pounds (4.3 billion euros). The issue price of stocks was set far lower than the actual market price so that investors were already raking in fabulous profits within the first few weeks.

Competition in supplying and distributing water is technically impossible because utilities have the character of a monopoly—only one company ever has control over a given region. The state regulatory agency Ofwat was therefore set up to regularly review water rates and make sure that companies worked efficiently and made needed investments in maintaining and modernizing facilities. Before the start of each new year, water companies must submit a calculation of expenses and the investment they plan to make. The agency then uses this as a basis for determining water rates. Not until several years had passed did it emerge that many companies had not been making the investment they were supposed to. At the same time, however, they had made profits from water rates based on the calculation of investment they had claimed they would make.

English water suppliers had for years paid out many billions of pounds in dividends to their stockholders while grossly neglecting to maintain the infrastructure. Thames Water alone, the water group responsible for London, had paid out no less than 2.3 billion pounds (3.8 billion euros) to its stockholders in the first ten years after privatization. This turned English waterworks into darlings of the stock market and made them attractive for takeovers as well. The German utility RWE in 2000 bought Thames Water for 7.1 billion euros, thus becoming the third-largest group in the international water business.

Much to the disappointment of RWE and the stock market, the bubble burst in Britain's water economy. It has now become painfully obvious that the high profits gained by privatized waterworks are missing for maintaining the infrastructure. The regulatory agency Ofwat is now putting pressure on companies to repair leaky water mains, many of which are well over a hundred years old. Because Thames Water neglected to maintain and repair mains for so many years, there are serious delivery bottlenecks in London. Not everyone living in London can take it for granted that the supply of drinking water is reliable. Thames Water lowered water pressure in its entire system to reduce the amount of water lost from leaking mains, but this means that the upper stories of high-rise buildings have very low water pressure. Tap water flows either very slowly or not at all, much to the anger of both residents and business people in office buildings.

More than one-third of the water Thames Water feeds from its reservoirs into mains is lost before it reaches households. According to official sources, 345 million cubic meters of expensively purified and sterile drinking water in 2002 seeped unused into the ground from leaky pipes. This is the same volume of water consumed by the 4.6 million people populating Berlin and the area around it.

To compensate for water loss from the mains network, Thames Water would like to build a seawater-desalination plant. The company argues that feeding in additional water from desalination is cheaper than repairing pipes. London's mayor didn't follow the logic of this argument and rejected the building permit requested for this project, deeming the loss of large amounts of water from a piping system in need of repair to be irresponsible.

Thames Water very hesitantly continues repairing pipes as work is expensive and extremely complicated due to years of neglect. The regulatory agency did pass a water-rate hike to finance repair work, but not by nearly as much as Thames Water had requested. This means that required investment will significantly reduce the company's profits over the next few years.

Ofwat and the population are angry, but it seems the weather gods are as well. An alarming supply crisis has been looming since late 2004. The south of England experienced its driest summer since 1933–for months, rainfall was just a fraction of what it normally was and Thames Water's reservoirs were only moderately full by early 2006. The year-long neglect of the mains network is now taking its toll. Although water reserves are becoming increasingly scarce, Thames Water still has to feed 35 to 40 percent more water into the system than reaches consumers. The population will have to bear the brunt and prepare to face shortages for months at a time. A lighter version of this scenario involves only a hose-pipe ban, but if worse comes to worse, it will involve supplying drinking water from water tankers and bottles.

Meanwhile, investors have learned from their experience with the private water sector. The financially strong utility RWE, which had enthusiastically bought into the water sector in 2000, was announcing late in 2005 that it intended to sell Thames Water as soon as possible because its expected return on investments could not be realized. It has become clear not only to Londoners that record profits from the domestic water sector of the 1990s were made largely because money was not being invested in taking proper care of the supply system. The capital paid out to stockholders will have to be raised once more from consumers, in the form of increased water rates, so that rundown facilities can function again.

Photo
London, Great Britain. Photo taken in 2002.
Peter Marlow/Magnum Photos

Laws of economics dictate that the scarcer a good is, the higher its price. So will water inevitably become more expensive too? Is water really the profitable and secure capital investment that many investment companies say it is? A number of arguments speak against this.

Is water a prime commodity for investment?

Equating water with gold or oil is unrealistic–water is simply not a commodity. Except as bottled mineral water, water is not traded around the world or even between regions. The weight of water alone, 1,000 kilograms per cubic meter, puts an ultimate limit to such activity. Transporting water over long distances would require extensive distribution networks across several countries or even between continents, a prohibitively expensive and impractical venture.

Indeed, practically nowhere on earth are rates charged for water itself as a resource. Consumers, whether households or businesses, pay only for the costs of abstracting, purifying, and distributing drinking water. Waterworks need more than 80 percent of the fees paid for water to build and maintain pipelines, pumping stations, and treatment facilities. The only commercial share in the price of water is the profit calculated in by the waterworks, as a rule significantly higher when water is supplied by private operators than by public ones.

The price we pay for our drinking water is not determined in a market for a scarce resource in high demand like gold or oil. Although turnover in the water sector amounts to billions of dollars each year, there is no actual business done with the physical resource of water and scarcity plays the most minor role in its pricing. If profits are made in the water economy, they are solely in constructing, maintaining, and operating the infrastructure.

Some investment bankers don't seem to possess this basic knowledge, judging at any rate by the glossy advertising brochures for some mutual funds invested in the water sector. Bank leaflets addressing financially strong investors suggest again and again that increasing demand and increasing scarcity will make water a highly lucrative object for speculative investment. One brochure claims that water is "your commodity investment," arguing that: "Given the predicted global demand

for water in the coming decade compared to supply, experts are already calling this commodity the 'gold of the future.' Dwindling resources and rising demand make water a precious commodity there is [sic]. No other natural commodity is experiencing such a rise in demand (annually 2–3 %), while resources are shrinking at the same time." (S&P Custom/ABN AMRO Total Return Water Index, Internet Offer 2006).

All current water funds (Bank Pictet, ABN AMRO, SAM Sustainable Water Fund) are composed mainly of stocks from the largest private water suppliers–Véolia, Suez, and RWE. Increased profits are mainly expected as a result of further privatization of the water sector. "Although just 38 % of the European population is supplied by private firms today, experts anticipate this figure to double to about 75 % by 2015. This rate is expected to more than quadruple in the USA, with the proportion supplied by private suppliers rising from 14 to 65 % in 2015." (ABN AMRO).

Fund managers understand very well the prime importance of using water efficiently. On the question of global water problems, the ABN AMRO brochure states in bold print: "The only solution: greater efficiency in the use of water in agriculture, industry and in drinking water processing and purification." But this important insight is hardly reflected in the representation of companies in the fund's portfolio–a mere 11 percent of the fund's capital is in stocks of businesses that look to improving efficiency.

Water funds are often described as ethical investments, allegedly allowing the investor to exert positive influence on living conditions in poorer countries. But the fact is that whoever invests in current water funds is doing very little or nothing at all for those 1.3 billion people around the world who don't have access to safe water. It is a proven fact that the large private water groups–making up the major share of all funds–are interested only in supplying water to wealthier cities with rapidly growing industries (see chapter on privatization). In recent years they have been withdrawing more and more from developing and emerging countries since the profits they seek there are not guaranteed. There is hardly any justification for categorizing water funds as an ethical investment option as some fund brokers do.

But what can private investors do to make a contribution towards alleviating the water crisis around the world? Especially if they would prefer investing money instead of donating it? There are two possibilities. The riskier one is to invest in businesses that work on technical innovation in the water sector–such as simple, water-saving irrigation technology for agriculture or production processes that avoid discharging wastewater (→ pp. 368). Some businesses already in the portfolios of water funds are possible candidates.

The second option is to give direct financial support to local authorities that are expanding their water supply services. To improve and expand water infrastructures, towns and cities issue bonds, borrowing investment money from national and international capital markets. Such fixed-interest municipal or government bonds are used to improve drinking-water supply and wastewater disposal and have the potential to improve the living conditions of thousands of people. Investors are paid a fixed rate of interest which is below what many banks promise their funds will deliver–but then often don't.

Photo
Poona, India. Photo taken in 2005.
Hollandse Hoogte/laif

Western science and technology are not always superior to traditions and customs that are hundreds of years old. An irrigation system practiced on Bali, closely linked to religious rites, has proven to be so successful that an attempt to modernize the irrigation of rice paddies was called off after a few years.

Bali's rice farmers and their water priests

The cultivation of rice on the Indonesian island of Bali boasts a long history. The oldest surviving inscriptions (896 AD) on the island refer to architects of irrigation tunnels. Small farmers on Bali's steep but fertile mountainsides of volcanic soil have to share a limited amount of water from small streams and rivulets with seasonally varying flow. Over centuries they have developed the *subak*, a cooperative arrangement for irrigation that is closely associated with religious values. About one hundred farmers join together in a *subak* cooperative. Each *subak* has a water temple and a water priest who, together with the farmers, decides how water will be distributed.

There are also temples and priests at all important branches in the irrigation system–they regulate the distribution of water from main canals to the various *subak*. The higher a temple is located in the water distribution system, the greater its importance. The most important and most sacred temple is Pura Ulun Danu Batur at the crater lake near the summit of the Batur volcano. Farmers make a pilgrimage here once a year to make a sacrifice to and thank the lake goddess Dewi Dano. Here a brotherhood of priests keeps watch over the distribution of water to about 1,300 *subak* and their rice fields. While priests are chosen by a medium in a trance and hence regarded with high esteem, they do not have any power or authority over the farmers.

Bali's traditional way of growing rice is sustainable and highly productive. In spite of

the tropical climate, it is also remarkably resistant to pests. One immediately noticeable feature is that all farmers on the entire mountain plant and harvest their fields, and leave them fallow, at the same time—which at first view seems absurd because they all need water at the same time.

In the early 1970s, this paradox also attracted the attention of Italian and Korean irrigation engineers entrusted by the Asian Development Bank, at the behest of the Indonesian government, to centrally organize irrigation in Indonesia and "modernize" Bali's *subak* system. The main idea behind this new agrarian policy was to use scarce water more efficiently through centralized planning and feed water to fields flexibly according to the needs of the day.

The foreign engineers found the *subak* water-distribution system to be arbitrary and recommended that it be disbanded and replaced by a rational system for measuring rain, river water volume, canal capacity, losses, and irrigation volume needed. Data was supposed to be recorded centrally so that engineers could calculate optimal distribution. But wherever the new system of staggered cultivation cycles was introduced, farmers were confronted with massive pest problems. Rats, brown leafhoppers, and other insects ate their way from field to field, reproducing uncontrollably. The *subak* system, in contrast, had kept these pests at bay—simultaneous cultiva-

tion cycles had eliminated their food sources because all fields had lain fallow at the same time. Even the use of pesticides dictated by the agrarian planners didn't have any lasting effect except to massively pollute local groundwater and drinking water with toxic residues.

The project was officially called off after a few years. Farmers had anyway on their own returned to their traditional *subak* associations. It was quite apparent that two incompatible ways of thinking had collided with each other in Bali. While agrarian engineers wanted to adapt the supply of irrigation water to the needs of plants, *subak* farmers had always taken into account the actual volume of water in the river, and the need to simultaneously plant fields to keep down the number of pests, to determine the best time for planting, and how intensive irrigation should be. Water priests ensure there is a connection between knowledge of the system as a whole and the amount of water expected. For the *subak* to function, it is important that farmers feel confident they are in harmony with natural cycles and can rely on the certainty of hundreds of years of experience gained by priests and farmers. It is Dewi Danu alone who holds her protecting hand over everything—not the market, nor the administration, nor the law.

Photo
Central Highlands, Bali, Indonesia. Photo taken in 2000. Stuart Franklin/Magnum Photos

Citizens' initiatives in India are becoming increasingly successful at battling international water groups that use cheap drinking water to manufacture expensive drinking water while massively polluting the environment in the process.

The women of Plachimada

Hindustan Coca-Cola Beverages, the Indian offshoot of the American soft-drink group Coca-Cola, will have disagreeable memories of August 19, 2005. On this Friday, environmental authorities in the Indian state of Kerala ordered the closure of a bottling plant in Plachimada, a little village in the district of Palakkad. Environmental authorities thus for the time being ended a three-year campaign of protest led by a committed group of women from Plachimada against the American corporation, a campaign which had generated a spectacular sensation throughout India.

It had all started with a happily mutual agreement made at the end of the 1990s, when Coca-Cola leased a 16-hectare plot of land in Plachimada to build a bottling plant for its soft drinks—Coca-Cola, Sprite, Fanta, Limca, Kinley Soda, and Maaza. Plachimada's local council in March 2000 granted the corporation a concession for harvesting groundwater by means of diesel pumps.

But unpleasant consequences were soon noticeable when Coca-Cola pumped up to 1.5 million liters of water per day from nine new well fields. Within three years, the water table in the surrounding area had

dropped from a depth of 45 meters down to 150 meters. Some 260 wells from which the women of Plachimada had previously drawn water ran dry. Where water was still to be had, its salt content and hardness had increased, and rice farmers recorded markedly lower crop yields—which Coca-Cola's lawyers later attributed to unusually low rainfall during these years.

But changes in the regional climate were not enough to explain other changes such as the extreme pollution of groundwater with chemicals. The BBC reported in July 2003 that water and soil samples from Plachimada analyzed at the University of Exeter contained extremely high concentrations of heavy metals, cadmium, and lead.

The party responsible was quickly found. The Coca-Cola factory had dumped its toxic waste in the open near the plant site. During the monsoon period, rain had washed toxins directly into irrigation channels and rice fields. When authorities put a halt to this practice, the toxic waste was dumped together with highly polluted industrial effluent into abandoned wells on the factory grounds, from where it seeped into groundwater. In an additional cunning move, the factory had sold foul-smelling and highly toxic sludge waste to farmers as fertilizer—later giving this waste away just to get rid of it.

In April 2003, after initial protests by the women of Plachimada, the local council annulled the Coca-Cola factory's three-year-old license. The highest court ordered a stop to production until the accusations against Coca-Cola could finally be properly investigated. In August, the environmental authorities of Kerala state confirmed that the water table had sunk and that groundwater was hugely polluted. Health authorities declared that the meager quantity of water in remaining wells was no longer suitable for drinking and could be used for irrigation only to a very limited extent. Coca-Cola defended itself with counter-arguments by experts who listed numerous other hypotheses to explain the scarcity of water and its pollution. The corporation described the protest demonstrations by the women of Plachimada as politically motivated, left-wing conspiracies. For a time, Coca-Cola was even able to lean on a high-level representative of the environmental authority who confirmed, despite knowing better, that the pollution from heavy metals was within tolerable limits. The official is now being investigated—he is said to have accepted bribes from Coca-Cola.

Observers surmise that the shutdown order issued by Kerala's environmental authority could have far-reaching consequences for a number of other factories producing bottled water and soft drinks. The problems created by the Coca-Cola factory in Plachimada when it extracted enormous quantities of groundwater, and especially the consequences for the local population, arise in other places too. Coca-Cola and Pepsi-Cola in India alone operate ninety bottling stations, each of which uses about the same volume of water—a total of about 40 billion liters per year. Producers of bottled water would like to greatly increase this number. In addition to Coca-Cola and Pepsi-Cola, the Indian market leader Parle Bisleri, the Swiss food group Nestlé, and other international corporations are planning to increase their capacities enormously. Not without reason—the deteriorating quality of drinking water in many parts of India promises huge rates of growth in bottled water. Vandana Shiva, a prominent Indian expert on water, estimates that market volume will double every two years in the near future. From 1992 to 2000 alone, the sale of bottled water and soft drinks in India increased tenfold, from 95 million to 932 million liters.

Photo
Popular protest against the exploitation of groundwater by Coca-Cola, Plachimada, Kerala, India. Photo taken January 2004. Klerx/laif

One of the most common arguments used in the debate on whether water utilities should be privatized is that water is used more sparingly only when consumers have to pay for it. But although Ireland, for instance, provides water at no cost to its population, the Irish have not become wasteful with it.

Beyond economy—the Irish experience with water pricing

"Anything scarce and in demand commands a price; this is one of the basic principles of economics." This sentence appeared in "Water Pricing," an article on the right price of water written in 2003 by Tom Jones, head of the global and structural policies division in the environment directorate at the Organisation for Economic Co-operation and Development (OECD). Jones was effectively transferring the economic principles of trade to dealings with water,

the elementary substance needed for life. In other words, only consumers who have to pay for water will use it carefully. This theory still continues to stand untested, unproven, and rarely questioned.

Ireland is a unique test case for this thesis. The Irish government has pursued a policy of its own, allowing everyone who lives in Ireland to be supplied with drinking water for free. The costs for waterworks and

pipelines, and for pumping and treating water, are met completely by the central government. Some of this money is recovered in turn from commercial users, be they hairdressers, butchers, or manufacturing industries, who all pay for water usage by the cubic meter. To be able to follow through with this policy, Ireland in the 1990s successfully fought against a cost-covering principle being made compulsory in the European Union's water framework directive.

According to the OECD's line of reasoning, people in Ireland should be using tap water at home profligately, since the typical *homo oeconomicus* can't appreciate anything it gets for free. But average consumption per capita per day is about 148 liters, hardly more than in model water-saving countries like the Netherlands and Germany (and less than in Austria and Sweden). This is the case in spite of the fact that water-saving fittings and toilets–important elements of efficient water use–are not even available for purchase in Ireland.

The Irish can rest assured that their country is blessed with plenty of rain and therefore with sufficient water reserves, hardly more than 2 percent of which are used. But at second glance, this picture becomes clouded–high-quality water reserves are becoming scarce in Ireland too because of rapidly increasing water pollution, exacerbated by the use of fertilizers in agriculture. At the same time, much more water is being taken from rivers and groundwater than these sources can cope with during dry years–especially in the metropolitan region around Dublin, Ireland's capital city.

So do the Irish use too much water? Indeed, waterworks abstract much more water from natural reserves than consumers let flow from their household taps. In Greater Dublin, waterworks feed around 500,000

cubic meters into pipelines each day. For a population of 1.3 million people, this amounts to 385 liters per person, more than double the actual consumption in households. Apart from the water used for commercial purposes, most of the difference is due to wastage. Authorities estimate that more than 40 percent of total water volume seeps from leaking pipes. Many water mains in Ireland are more than a hundred years old and should have been replaced long ago.

If good water does become scarce in Ireland one day, false economics will not be to blame. Supplying water for free has not caused people to rampantly squander it. It can be assumed that campaigns for saving water, which have yet to take place in Ireland, would reduce household consumption even more, especially if retailers and plumbers were to make available the water-saving fittings and dual-flush toilets that have been standard equipment in other countries for a long time now.

There is however an urgent need to look after the infrastructure. Irish water reserves are overused mainly because of a massive loss of water from leaky pipes. This is not a problem of economics but an issue of inadequate control by regulatory agencies and authorities.

Many citizens fear that the Irish government could one day bow to the pressure of OECD ideology and abandon its policy of providing water for free. These developments are carefully observed by international water-supply corporations who currently play no role in Ireland because of this very policy. The water industry has long been among the most ardent proponents of the principle of full-cost recovery, not least because it benefits them directly. The very principle that should motivate people to use water carefully at the same time guarantees flourishing business in a privately run water sector. It is at the core of the principle of full-cost recovery that all costs incurred by the operators of waterworks have to be covered by consumers. No matter how efficiently or inefficiently an operator works, no matter what quality of service it provides, this principle always ensures handsome profits.

Photo
Harry Gruyaert/Magnum Photos

For a long time, established science believed that traditional wisdom handed down by woodcutters was pure superstition. Tradition had it that the quality of wood changed as the moon went through its phases. But in recent years, several experiments have proven there is a connection. Scientists are beginning to revise their thinking.

Of moon, wood, and water

If trees didn't seem so ordinary and natural to us, we would consider them miracles. How do nutrients and moisture in the soil actually get all the way up to the crown of a tree, often 70 or 80 meters above the ground? If we wanted to mechanically construct such a head of water, we would have to use powerful pumps.

The upward movement of water in trees is driven solely by evaporation from the surfaces of leaves. The negative pressure in tree trunks created by leaf transpiration sucks up water from the soil through the roots. Water is not pushed from below but pulled from above–against its own weight. Because of water's inner cohesion, a thread of water is extremely strong, even if it is only a millionth of a millimeter thick. Recent research into water transportation through cell membranes has shown that threads of water as thin as a single molecule still behave like a real liquid.

Water also plays a decisive role in determining the properties of wood. A traditional belief handed down for centuries is that the quality of wood depends on when a tree was felled. The weather and the season are important, but tradition has it that lunar phases are important too (the change from new to full moon and back, with quarters in between). The wood of trees cut down during the new moon, it is said, is more resistant to fire, pests, and fungi. Wood chopped down before the full moon is said to be good firewood. Depending on how wood will be used, the optimal time for cutting down a tree is sometimes linked to the height of the lunar orbit in the night sky, its relative changes (ascending and descending lunar orbits), or to the position of the moon in relation to constellations in the zodiac.

Such age-old rules regarding the right time for felling trees, sometimes handed down only through oral tradition, are found in many places–the Alps, the Middle East, Africa, India, Sri Lanka, Brazil, and Yucatan. Many scientists believe that any influence of the moon on wood quality is unlikely.

Today's understanding of the physics of wood does not offer any plausible explanation for interaction since the gravitational pull of the moon is too weak to be the direct cause of the effects observed. But traditional descriptions of variations in the characteristics of wood, seen in relation to the moon, are so astonishing that a number of research teams have been trying for years to find a possible cause of these effects.

A SLOW AND GENTLE WAY
TO LET WATER EVAPORATE
Trees are cut down on special, favorable days. Moreover, they are occasionally felled in a downhill direction and branches are sawn off later, sometimes not until after snow melts. What effect the direction trees are felled in has, and what happens in stems during the days after they are cut down, isn't yet clear. But it is certain that a felled tree continues to transpire moisture and that this process is intensified when spruces and firs, for instance, are felled in a downhill direction so that water can flow towards the tree's crown along the grain of the wood. This gentle and natural drying process is much more beneficial for wood quality than when water is forcibly removed in kiln dryers.

Experiments demonstrate beyond any doubt the differences between new-moon and full-moon wood. Such experiments must take into account the time of felling, the location and condition of the tree, and the position of the samples in the trunk and their direction, for findings to be as accurate and repeatable as possible. Differences in freshly felled wood are negligible. The typical characteristics of new-moon and full-moon wood do not develop until the wood starts to dry, in other words, when water evaporates.

Differences are evidently related to the wood's water content. There are two different forms: water which freely circulates in cell cavities, and bound water in cell walls. Free water evaporates more easily, the wood dries more quickly and shrinks (loses volume) very little. In contrast, water bound in the cell walls evaporates more slowly and to a lesser degree, and needs more energy to do so. The bulk of the wood shrinks more and wood density significantly increases.

Tests show that trees felled on certain days during the waning lunar phase lose less water but shrink more. This results in denser and harder wood, qualities that are particularly desirable in timber used for building and construction. Its resistance to pressure is higher too, possibly indicating better resistance to attack from fungi and insects. These findings tally with the old rule that new-moon wood is more durable. In contrast, it is easier to remove water from trees felled on certain days of the waxing lunar phase, thus increasing fuel value. Full-moon wood tends to be lighter and shrinks less, possibly explaining why to this day it is the preferred wood (resonance wood) for making musical instruments.

Does full-moon wood contain more free water, while water in new-moon wood tends to be bound in cell walls, and if so, why? Which factors determine these phenomena? Does the lunar phase play a role, or the terrestrial magnetic field, or the periodically oscillating electrical field of Earth's atmosphere?

What is perplexing about the physics of wood is that the relationship between free and bound water periodically changes–not only in the living part of the trunk, its peripheral, relatively thin sapwood, but in its heartwood as well. The current state of technology for processing wood assumes that the characteristics of heartwood remain stable over time. But it is now becoming evident that the strength of the wood-water bond, even in heartwood, varies with the changing of the seasons and lunar phases, just like it does in living sapwood.

Sensitive electrodes attached to trees are moreover able to measure bioelectrical currents that indicate variations in the monthly and daily rhythm of the moon, particularly in winter months. In spring and summer, when temperature and humidity vary depending on whether it is night or day, and trees experience strong growth, vegetative sap flow is superimposed on the lunar rhythm. But in late fall and winter, when trees rest, the basic rhythm of the moon is decisive–certainly the reason why ancient woodcutting wisdom regards the winter months, in combination with cycles of the moon, as the best time to cut down trees.

A CONCLUSIVE THEORY
ON THE MOON'S INFLUENCE?
Only more recent research can possibly explain why water can be removed from trees with varying degrees of ease depending on daily and monthly rhythms. Scientists recently proposed a seminally new geophysical model in the form of quantized gravitational force, resulting in the conclusion that the supramolecular aggregation of water–clusters of coherent molecules as

it were—varies with the rhythm of sun-earth interaction, and at the same time, with the rhythm of moon-earth interaction.

By coincidence, the architect of this theory noticed that fluctuations in the properties of water, as predicted by computation, were accurately mirrored in the behavior of trees. The diameter of young spruce trees growing under unchanging conditions increases and decreases rhythmically with lunar phases and tides. Until now, wood researchers have explained that pulsation in the thickness of trunks was due to water periodically moving between cell cavities and wood cells. Since this pulsation takes place in exact step with the rhythmical change in water's properties as computed by physicists, the effects of the moon on wood could also be explained by water's periodically changing degree of aggregation.

The model of quantized gravitational force means that water in biological cells builds molecular clusters of varying size depending on the periodically changing gravitational pull of the moon. The properties of water, including its tendency to evaporate, would depend on the size of its molecular clusters. If this theory proves to be correct,

theoretical physics may well have found an explanation for the phenomena observed in trees for centuries—as well as the agent for many other lunar interactions with plants, animals, and humans. Physicists and wood biologists, running experiments to identify the state of water in wood, are hard at work to test the predictions derived from theoretical models.

Literature
– Dorda, G.: Sun, Earth, Moon: the Influence of Gravity on the Development of Organic Structures; Part II: The Influence of the Moon. München 2004.
– Holzknecht, K., 2002: Elektrische Potentiale im Splintholz von Fichte und Zirbe im Zusammenhang mit Klima und Mondphasen. Diss. Nr. G0643, Inst. für Botanik, Universität Innsbruck.
– Zürcher, E., Cantiani, M.-G., Sorbetti-Guerri, F. and Michel, D. (1998): Tree stem diameters fluctuate with tide. Nature (392): 665-666.
– Zürcher, E. (2000): Mondbezogene Traditionen in der Forstwirtschaft und Phänomene in der Baumbiologie. Schweizerische Zeitschrift für Forstwesen 151 (2000) 11: 417-424.
– Zürcher, E. (2003): Trocknungs- und Witterungsverhalten von mondphasengefälltem Fichtenholz (*Picea abies* Karst.). Schweizerische Zeitschrift für Forstwesen 154 (2003) 9: 351-359.

Photo
Before moon wood is felled, a woodsman examines the state of the trunk and determines which way it is to fall. Trittau near Hamburg. Michael Kottmeier/agenda

The machinery of politics turns slowly. Sluggish state administrations, bureaucracy, and powerful interest groups often prevent the application of rapid solutions to even the most urgent of problems. Pressure from the public frequently helps to get political processes started. Today, non-governmental organizations, citizens' initiatives, aid agencies, churches, unions, and other associations in civil society play a more and more important role—not least in political decisions regarding water.

Water, power, and NGOs

Political systems generally resist change. In parliamentary democracies, new political ideas are normally worked out first in the programs of political parties and are only slowly implemented in practice. The management of water—enforced by administrative rulings that are often principled if not downright rigid—is traditionally determined by the most conservative of policies. But in recent years, water management has become much more flexible.

There are two reasons for this. First, the pollution of rivers and lakes became more

visible during the second half of the 20th century. The press in western industrialized countries had to report on rivers that stank, dead fish floating on their surfaces together with mountains of foam from detergents. Swimming was banned at lakes and seacoasts, acid rain fell from the sky, and wells dried up. Poor water management in the 1970s and 1980s could no longer be ignored. Disasters like the fire at a Sandoz chemicals warehouse near Basel in 1986 (→ pp.302), which destroyed all life in the Rhine River, forced politicians to act. Second, in reaction to these circumstances,

activist groups and organizations that dealt with environmental and water-management issues were founded. These independent non-governmental organizations (NGOs) brought fresh ideas to the arena of water policy. They created awareness of the pollution and overuse of bodies of water, documenting damage and who had caused it and drawing attention to alternatives and solutions. Environmental organizations like Greenpeace, the World Wide Fund for Nature, and Friends of the Earth, as well as national groups like the Sierra Club in the United States, from the very start focused on protecting natural waters and drinking water.

At first, these pressure groups met with fierce resistance from administrative authorities and the companies responsible for pollution, mostly chemicals businesses, paper manufacturers, and other industrial groups. The establishment didn't want to give up its monopoly of power over water to "self-appointed environmental evangelists." It wasn't until environmental organizations gained in credibility and were supported more and more by the public that fronts softened. Even the conservative community of waterworks operators joined the ranks of environmentalists when they realized that protecting rivers and groundwater would also prevent harmful chemicals from entering the drinking-water reserves they managed.

At the same time, political initiatives focusing on development and combating poverty also turned more attention towards water management. The groups involved have called for all human beings to have reliable access to good drinking water and for the basic human right to water. Church groups, charitable aid associations, and labor unions play an important role here. Environmental and development organizations not only have a place in discussions on international water policy, they also launch specific aid projects. They argue with the World Bank and with governments over the direc-

tion of state aid programs and they exert influence on the drafting of national water laws and international conventions. At the same time, they use donations from the public to improve access to drinking water and sanitation facilities for millions of people.

What began in the 1970s and 1980s as a handful of political activists has in the early 21st century become a worldwide, highly organized movement of NGOs, an important regulator of societal and political processes. NGOs have in the meantime become a solid component of political life, even in most developing countries. This is where globalization works–NGOs from north and south work together and find rapid and uncomplicated solutions that would hardly be possible between governments. The exchange of information through e-mails and the Internet leads to an accelerated process of learning. Knowledge of specific political circumstances or options offered by technology is available everywhere at all times.

NGOs today exert noticeable influence on political decisions regarding water issues. Pollution, privatization, dam construction, a human right to water–all these are issues where NGOs not only actively join the discussion but from time to time have the greatest say. This influence goes too far for some governments and businesses who want to keep civil-society groups out of decision-making processes. Those who want to continue preparing and making decisions out of the public eye indeed feel NGOs restrict their power. Critical voices have already brought about the downfall of contentious decisions or basically changed them. This is not just more democratic, it's something even better–decisions made with public participation are more objectively balanced and far better accepted by the citizens they affect.

Thanks to NGOs, the voice of civil society is heard unmistakably at major events where water policy is determined, like the 2006 World Water Forum in Mexico, where experts, development banks, industry, and governments debate on strategic decisions. Although representatives from NGOs are allowed only rarely into sessions and consultations, it is their media presence that influences issues and results. Official parties can no longer make agreements behind closed doors, as they still did just a few years ago, without taking public opinion into account. The counterweight of NGO expertise has made it much more difficult for governments and big business to assert their special interests. This makes the NGO community a monitor of global society that creates transparency and works as a whistle-blower warning against one-sided or shortsighted decisions.

The way in which a society deals with water issues is not a constant. Politics and culture are subject to a continuously unfolding process. Every stimulus, every talk, every action basically contributes to culture and politics. Through the experience that change is possible, more and more people develop from subjects to citizens, to active subjects who see their surroundings and the world as an entity that can be influenced. The exertion of influence leads to the sharing of responsibility and creates a feeling of belonging. The creative spirit of change can open up wherever people no long feel they are suffering or are victims, wherever they can exert influence.

Social involvement is sensible and personally satisfying. People can join existing NGOs or found their own interest groups. All important organizations have websites. In recent years, petitions for referenda and plebiscites have become options in more and more countries and cities, and in many places people have launched initiatives against the privatization of their own local

water supply. In Hamburg, Germany's second-largest city, a group of active citizens got together in February 2003 to block the plans of the city government to privatize the city's water utility. Supported by local and national NGOs, the group launched a petition for a referendum to keep the city's water supply in public hands. Some 150,000 people signed the petition in the summer of 2004. As a result, the Hamburg parliament called on the city government to draft a law to reflect this interest. The Hamburg law on keeping the water supply in public hands, passed in 2006, is unprecedented in Germany, and thus Hamburg is the first city to forbid the sale of its public water utility or to allow private operators to manage the water supply.

There are many ways in which citizens in a free society can make themselves heard. We should use this privilege to remind those in power of their responsibilities regarding water. At the same time, we should not forget what we can do ourselves where water is concerned. A respectful attitude towards water creates the basis on which we can use our own ideas to publicly influence water management.

Photo
Cumbria, Great Britain. Smith/Greenpeace

Bangladesh has succeeded within a few years in greatly improving the supply of latrines to its population, thanks to intelligent policies and well-designed campaigns. By 2010, some 150 million people should have access to a latrine. There is a good chance that the government will actually achieve this ambitious goal.

The Bangladesh sanitation miracle

Barely twenty years ago, the southern Asian country of Bangladesh was facing an almost insurmountable problem with hygiene. Not even 10 percent of the population had regular facilities for defecation, and more than 90 percent of the 100 million inhabitants of Bangladesh at that time had to relieve themselves in the open because there weren't enough latrines. As a consequence, hundreds of thousands of people suffered from diarrhea and child mortality was shockingly high.

The government attempted to meet this challenge with a state-subsidized program to build latrines. About 600 state latrine factories produced a standard-unit latrine that was very cheap because it was highly subsidized. Nevertheless, no need arose in practice for such latrines. When people did actually own one, they generally didn't use them for their designated purpose but rather as a storeroom or goat shed. Production capacity in state factories remained so low due to weak demand that it would have taken one hundred years to supply the

entire, rapidly-growing population. Many experts came to the conclusion that there was practically no demand for latrines in Bangladesh and that Bangladeshis were simply not prepared to pay for them.

In 1989, the government expanded its offer of subsidized latrines by initiating a demand-driven approach. Instead of trying to forcibly improve sanitation conditions with its offer of cheap latrines, it encouraged competitiveness and set up educational campaigns to create the desire in the population for better sanitation, at the same time opening a market for private latrine manufacturers. It soon emerged that the idea that Bangladeshis didn't want to spend money on latrines had been a false conclusion—albeit they didn't want to spend their money to prevent diarrhea (what use is a latrine if neighbors are still relieving themselves in the open). Indeed, women in particular appreciated the comfort and privacy of a latrine, and finally, prestige played an important role for men. Women and children experienced the advantage of having their own latrines most directly because they didn't have to stumble across fields in the dark looking for an unused place.

Bangladesh retracted the idea of the standard-unit latrine. Ten different models were developed, from cheap latrines people could build themselves, costing half a dollar, to luxury models with double pits and attractive housing for 20 dollars. This allowed customers—the heads of households—to choose the most suitable version from a number of models depending on their needs, personal taste, and especially their budgets. Economists point out that the state's withdrawal from the project was a decisive step because it left the marketing of latrines primarily to manufacturers and salespersons. Gradually, more and more latrine workshops were opened and by 1995, with more than 4,000 of them in existence, they had become a significant rural industry. Privately-run workshops were so successful that they crowded government workshops out of the business, even though the latter were still profiting from state subsidies.

Additionally, demand was raised by a social mobilization campaign initiated by the government, UNICEF, and various NGOs. This activated village chiefs, politicians, and imams. During this campaign from 1993 to 1998, the percentage of the population supplied with latrines rose from 15 percent to about 43 percent. A new campaign was launched in 2003, the Total Sanitation Campaign, for supplying latrines everywhere. Attention is now being focused on entire villages with the goal of introducing latrines for all and completely ending defecation in the open. Easy-to-remember participatory methods were developed to meet this goal. For instance, calculations are made at village meetings to demonstrate how much excreta a village of 150 families produces each year—and how much of that is deposited in the open if only 15 households have latrines. If 485 people excrete 800 grams of feces every day, after a year more than

140 tons end up in the open, enough to fill 26 small trucks. After villagers have all talked about this and what happens to this mountain of excreta, they generally decide to take action. The social pressure exerted in villages to ban defecation in the open, particularly to safeguard children's health, has strongly increased the demand for latrines. The government has declared a goal of supplying latrines to all 150 million Bangladeshis by 2010–five years earlier than the target set in the millennium goals. This seems to be realistic in view of the current political will. Demand for latrines is enormous in any case. Today there are at least 10,000, maybe 20,000 latrine workshops in Bangladesh. Altogether they can manufacture between three to six million latrines every year.

Source
Urs Heierli, *How to make sanitation work as a business* (Bern, 2006).

Photo
Latrine workshop in Bangladesh.
Urs Heierli

What can be done when nature plays tricks on governments? The divide between the water catchment areas of Chile and Argentina is buried under glacial ice in the border region of southern Patagonia. Although nearly impossible to delineate, this divide is also supposed to mark the official boundary between the two countries. The indiscernible line was a point of contention for decades–until an odd episode smoothed the way to an agreement.

An icy divide

A broad expanse of glaciers, a typical continental ice field, lies at the southernmost tip of South America in cold and inhospitable southern Patagonia. The border between Chile and Argentina runs through the middle of this icy desert. Chileans call it *campos de hielos*, but it appears as *hielos continentales* on Argentinean maps. The different choice of names for this remote and rough ice cap in the Patagonian Andes reflects the territorial claims made by these neighboring countries. The question of who possesses how much ice, and where the border precisely runs, was resolved only very recently after being a source of conflict for more than a century.

Both countries agreed in a contract signed in 1893 that the highest peaks parting the waters would determine the border, with Argentina calling all regions to the east its own, and Chilean territory lying to the west of this line. In other words, the divide was equated with the major crest of the Andes. Not until later was it evident that the crest and the water divide failed to match over many kilometers. The line of the watershed was particularly blurred in the region where glaciers flowed into the Pacific, the continental ice field.

Historically, territorial conflicts around lakes, rivers, and glaciers have proven to be very long-standing. Argentina and Chile nearly went to war in 1978 over three islands in Patagonia's Beagle Channel. This conflict was averted only through papal mediation. In much the same way, all attempts to delineate the border in the continental ice field failed. But the conflict was finally resolved when the Chilean ex-dictator Augusto Pinochet was unexpectedly arrested in Britain in October 1998.

Pinochet's house arrest in London by court order provoked a vehement anti-colonial

reaction in South America. Carlos Menem, president of Argentina at that time, took a stand in favor of Chile, condemning Pinochet's arrest as an illegal action by Spain and the United Kingdom. At the next Mercosur summit in Rio de Janeiro, six South American presidents made a joint political declaration in which they fundamentally rejected the right of European judges to pass sentence on Pinochet.

This highly charged political intermezzo around Pinochet led to a softening of political tensions between Argentina and Chile. The Chilean foreign minister even publicly apologized for Chilean espionage favoring Britain against Argentina during the Falklands war. This gesture put pressure on Chile's government and parliament to give

up their resistance to a treaty delineating the border through the continental ice field. At the same time the border was being drawn in the ice field, important water resources were also allocated to each country by mutual agreement. All streams and rivers in the southern Patagonian Andes flowing into the Río Santa Cruz were allocated to Argentina and all waters draining to the western fjords on the Pacific coast to Chile. This forward-looking decision is an important contribution towards understanding and friendship between the two countries. But it would hardly have happened if it hadn't been for Argentina's categorical rejection of Pinochet's arrest.

Photo
The Perito Moreno Glacier, an ice-field on the border of Chile and Argentina. Olivier Leupin

New Orleans for a long time faced an ever-growing threat of disastrous flooding caused by powerful hurricane winds. But after decades of relative quiet, the city had lost its sense of apprehension.

New Orleans—a disaster waiting to happen

Millions of people around the world recently stared at their television screens in bewilderment, watching New Orleans become inundated with water while thousands of people were stranded in the heat at a big sports stadium, without enough water or food, pleading for help and crying futilely to be rescued. People shooting and looting,

houses burning, chaos reigning supreme. American television announcers were shocked to discover that even days after Hurricane Katrina swept waves more than 10 meters high over the city before finally dissipating, the situation couldn't be brought under control.

DRAINAGE LOWERED
THE FOUNDATION

Apocalyptic scenes of New Orleans could give the impression that the city was destroyed by winds and flooding of such unprecedented power and height that even superpower United States was unable to withstand them. But that's not how it was. The disaster in New Orleans was pre-dictable and experts had repeatedly issued warnings in recent years. The question why it couldn't have been prevented anyway and why authorities failed so spectacularly at coping with it will certainly preoccupy the United States for a long time. It was beyond doubt that the threat of disastrous flooding in Louisiana's Big Easy on the Gulf of Mexico was increasing. There are a number of reasons for this.

New Orleans, founded in 1718 by French explorers, was built on a small elevation in the swamps of the Mississippi Delta. Napoleon sold it to the United States in 1803. Today we know that most of the delta only accumulated at the mouth of the river during the past few thousand years. It con-sists of loosely deposited alluvial material that settles slowly, often under the weight of fresh deposits. The original city, now known as the French Quarter, was laid out in a sick-le-shaped area between a bend in the river and Lake Pontchartrain, which connects to the Gulf of Mexico.

Sinking ground and flooding were a prob-lem from early on. The New Orleans Sew-erage and Water Board in 1899 commis-sioned Albert Baldwin Wood, an inventor and city engineer, to improve the city's drainage system. Wood developed hydrau-lic facilities, most notably powerful pumps that needed little maintenance, and began at the beginning of the 20th century to dry out the swampland surrounding the city. This allowed the city to grow. Wood became a leading figure in American hydraulic engineering and also worked on projects abroad, among them the draining of the Zuider Zee in the Netherlands.

But New Orleans itself continued to sink more and more as drainage compacted its foundations; more and more massive and higher levees had to be built to seal the city off from Lake Pontchartrain and the Missis-sippi. The city lay like a bowl, its rim practi-cally level with water in the lake and the riv-er, and some parts of it as much as 2 meters below sea level, protected only by meter-high levees. These were built to withstand Category 3 hurricanes with flood surges up to 3.8 meters in height. But there had always been much more powerful hurricanes–the Category 4 Chenier Caminanda hurricane in 1893 swept over Louisiana and killed 2,000 people. Other human activities and their consequences had also reduced the cushioning effect that offshore islands on the delta exerted on hurricane waves. An article in *Scientific American* in October 2001 described a "drowning New Orleans" that could sink under 6 meters of water in a big hurricane, leading to the death of thousands of people.

PROTECTIVE SWAMPS
BEGAN TO DISAPPEAR

Swamplands decreased in area. But marsh-es are useful, experts tell us, in absorbing large volumes of water, thereby reducing the height of wave surges. The channeling of the Mississippi, carried out more than a hundred years ago, contributes to this fatal shrinking. It prevents new sediment from washing into the delta during flooding. This would deposit new ground or replace mate-rial that has been washed away. Instead, sediments are rapidly transported forward to the mouth and catapulted into the ocean where they disappear "unused" into the deep. This also means that fresh supplies of sediment never reach the offshore islands heavily eroded by constant wave action. When intact, these islands serve as break-waters.

Swamps and marshlands are also exposed to other corroding influences. The *Scientif-ic American* article pointed out that at Port Fourchon, a center of the oil industry, numerous channels have been excavated for ships and pipelines. Together with shipping traffic and tides, these help destroy marshlands. Added to that, saltwa-ter encroaches more and more on land farther inward, causing vegetation to die out, which in turn accelerates the ocean's advance even more. Experts reckon that some 70 square kilometers of swampland in the Mississippi Delta disappear every year, thereby reducing a protective buffer against hurricane waves.

Computer simulations at Louisiana State University came to the conclusion that evacuation would be impossible because the few escape routes would be cut off by flooding. The *Scientific American* article, written four years before the Katrina disas-ter, warned that more than 100,000 people could fall victim to such an event. Only a big project to restore the delta landscape could save the city. Such a project would entail shutting down old canals, building new canals and large flood gates to separate Lake Pontchartrain from the Gulf of Mexico during storms, and moving sand to reinforce the offshore islands. But it wasn't until experiencing the shock of Hurricane Georges, which rushed towards New Orleans in September 1998 with wave surges more than 5 meters high, changing its direction only at the last moment so that the wave front collapsed, that experts at different institutions agreed on a joint plan called Coast 2050.

FINANCIAL PROBLEMS

These plans were greatly reduced for cost reasons; it seemed that little money came from Washington for projects run by the U.S. Army Corps of Engineers either. The Corps, involved most notably in the development and maintenance of river canals–and present for more than 200 years in New Orleans–played a central role in flood protection. On May 23, 2005, a posting on its website indicated that important problems could not be solved due to the lack of money. The Corps was referring to a project that in the foreseeable future was to protect people living between Lake Pontchartrain and the Mississippi from wave surges caused by Category 3 hurricanes. Although Congress had raised the 3.9 million dollars originally proposed in the President's 2005 budget to 5.5 million dollars, the project needed 20 million dollars.

But shortly after Katrina struck, the Corps was saying that budget reductions played no role in the disaster, and that not even more rapid restoration of marshlands would have helped since wave surges came from the east. And besides, the flow of financial resources had remained constant. Moreover, it hadn't been a levee that had breached but a flood wall that had been overtopped with water which undermined its foundation (→ pp. 405). There hadn't been any plans to make improvements in this section. The whole system was designed to withstand a Category 3 hurricane, with a 0.5 percent probability per year of occurrence. Katrina was graded down from 5 to 4 shortly before it reached the coast. According to the Corps, said the *New York Times*, protecting New Orleans from a Category 5 hurricane would have cost 2.5 billion dollars. Washington is now talking about making available more than 10 billion dollars to repair Katrina's damage.

Erik Pasche, a specialist in hydraulic engineering at Hamburg-Harburg Technical University, is rather surprised that authorities in New Orleans did not chose more extreme events on which to base their calculations for protective measures. Nevertheless, he believes that not every flood can be prevented and that it's important to "give in to the water" during disastrous events, for instance, by following through on a meticulous plan for mandatory evacuation of the population. There are plans in Hamburg to set up an interactive website with detailed geographical data. The website, which will post precise information relevant to different sections of the city, is meant to raise awareness of the dangers of flooding among planners and the population.

There were plans to evacuate the population in New Orleans, too. Press reports related that some inhabitants were not prepared to leave the city again in long lines of cars after having fled unnecessarily from Hurricane Ivan in September 2004 and Hurricane Dennis in July 2005. At least the city had been in luck ever since 1965, when Category 3 Hurricane Betsy killed more than fifty people and flooded large parts of the city. In other places too, big disasters had to happen first before people took prevention seriously. In the Netherlands, for instance, new levees and weirs were built after nearly 2,000 people died in 1953 when dikes breached. In Bangladesh, where in 1970 more than 300,000 people died as a consequence of tidal-wave flooding caused by a cyclone, there are now high-lying protective areas that safeguard survival for many. And in Germany, flooding in recent years has made the population more aware of these events. It seems certain that New Orleans will start a new phase in its battle against flooding.

Translation of a text by
Heidi Blattmann

New Orleans still at risk

How great is the threat that New Orleans will again suffer the same fate it did in 2005? Ivor van Heerden, professor at Louisiana State University and vice-director of its hurricane research center, warns against underestimating the risks. In spite of repairs, levees offer no more protection today than they did before Katrina, far from it. He believes there would be another flood if a storm of the same category hit again. Even a moderate tropical storm would put parts of the city under water because pumps are now working at low capacity only. Van Heerden suspects that the soil under the flood walls is so unstable in some places that walls could collapse during the next flood. He is critical of the fact that sandy soil was used during rapid reconstruction–it would quickly erode if the situation turned critical.

Born in South Africa, van Heerden has a reputation for being outspoken. After years of warning the city in vain of a disaster like Katrina, he no longer hesitates to criticize authorities. He is particularly critical of the U.S. Army Corps of Engineers, the federal agency responsible for flood protection. For months the Corps claimed that the flooding had occurred because flood walls at two city canals had not been built to withstand such high water levels. They were allegedly overtopped and landside erosion caused their collapse. But van Heerden quickly came to the conclusion that water had never reached the top of the walls and

therefore faulty design and inadequate construction were to blame. He was right. Today, it is generally agreed that the Army Corp's poor engineering was at fault.

UNSTABLE SUBSOIL
A report published in 2006 by a research group at the University of California at Berkeley came to similar conclusions. Engineers incorrectly estimated the integrity of the subsoil. To save space, earthen levees, which were not high enough, were heaped up and topped with a cordon of concrete revetments. But these flood walls weren't sufficiently anchored in the unstable ground, allowing them to tip over under the pressure of floodwaters. Moreover, the metal plates they rested on hadn't been rammed into the ground deeply enough. In some places, water seeped under them, waterlogging the walls and causing them to collapse.

The research group calculated that three breaches at two canals were responsible for 80 percent of the flooding in the center of the city. This raises the question of why more attention wasn't devoted earlier to these weak spots. Two canals, one on 17th Street and the other on London Avenue, usually serve to drain the lower-lying parts of the city. After downpours, the water is pumped to the canals from where it flows into Lake Pontchartrain. But if the lake is flooded, which usually happens during a

hurricane, the flow is reversed and canals transport floodwaters into the heart of the city.

The obvious solution to this problem, placing a lock gate at the mouth of the canal, which could have closed during flooding, was never realized. For years, the two local offices responsible for maintaining and operating canals refused to carry out such a project. The Corps of Engineers turned to a makeshift solution, the slipshod construction of flood walls. Berkeley researchers see this as a good example of how poor cooperation between different government authorities can have disastrous consequences. They are calling for changes in engineering as well as institutional reform.

Katrina had to happen before bureaucratic resistance broke down. The Army Corps of Engineers has now been given the go-ahead to build lock gates. They won't be in place before the next hurricane season starts. But a visit to the 17th Street Canal proved that work on a gate there is progressing at top speed, and will presumably be finished by July 2006. If flood waves threaten the city earlier, a makeshift construction of metal plates will close off the mouth of the canal. Just a few steps towards the center of the city, past ruined houses, the place where the levee collapsed in August 2005 is still very visible. The breach has been repaired and temporarily stabilized with a coffer dam, a steel barrier. Although reconstruction of the

flood wall here hasn't started yet, a kilometer-long flood wall in the eastern part of the city has been completed on time. Construction methods were changed so that individual concrete elements are now more securely anchored in the soil. This provides better protection for the Lower Ninth Ward, which was flooded by Hurricane Katrina and again by Hurricane Rita.

DOES NEW ORLEANS HAVE A FUTURE?
The Army Corps of Engineers and federal authorities have admitted that a new Katrina would bring floodwaters to New Orleans again. It is expected that that levees would be overtopped once more and some of them could burst. But thanks to the sealing off of drainage canals, such massive torrents of water probably wouldn't pour into the city again.

Not that this is much of a consolation to hurricane researcher van Heerden. He doesn't understand how the United States can leave one of its most beautiful and historical cities in such a vulnerable state. He notes that Katrina was by no means "the big one," the dreaded storm of the century. Katrina did not roar directly over New Orleans and had already lost a lot of its force beforehand. According to van Heerden, the Louisiana coast is struck by a hurricane of this size about every seven years on average, and each storm could spell disaster for New Orleans. In his book *The Storm*, he calls on the public to refuse to accept stopgap measures. Constructing a levee system to protect the city and its surroundings even from the most powerful hurricane is feasible. If the city were to implement such a plan, every neighborhood could be resettled and people wouldn't have to be evacuated during future hurricanes.

Van Heerden describes his plan, which calls for constructing new levees and building a lock gate in the strait connecting Lake Pontchartrain to the Gulf of Mexico. Whenever floods threatened, the gate would be shut to protect the region around the lake from floodwaters. He sees as a model the engineering practiced by the Dutch, long familiar with such structures. He also makes a case for systematically regenerating marshlands located south of New Orleans. Over the last decades, many swamps have disappeared because of erosion, human intervention and the channeling of the Mississippi, which can no longer freely deposit its sediments. This has caused southern Louisiana to lose a natural buffer against storm tides. As marshlands disappear, the coastline continues to move dangerously close to New Orleans.

What would such an enormous project cost? Van Heerden estimates about 30 billion dollars. This is a lot more than the 3.3 billion dollars Congress approved for repairing the levees. But given the expense of the war in Iraq, costing some 8 billion dollars per month, or the cost of the damage caused by Katrina, estimated at between 100 to 150 billion dollars, the hurricane expert believes this is not asking too much. Katrina is a shame to the nation and should stir America's conscience. If nothing changes, he believes that at some time about one-fifth of Louisiana will sink beneath the waves forever. His dark warning: all that will be left of New Orleans will be the memory of a second Atlantis.

Translation of a text
by Andreas Rüesch

Photo I
Houses close to the Industrial Canal. Its levees were burst by the tidal waves caused by Hurricane Katrina. Photo taken on August 31, 2005.
Eric Gay/Keystone

Photo II
New Orleans, Louisiana, photo taken on 3 September 2005.
Thomas Dworczak/Magnum Photos

Photo III
Looters have set fire to houses in the Garden District. New Orleans, Louisiana, USA. Photo taken on September 4, 2005.
Thomas Dworzak/Magnum Photos

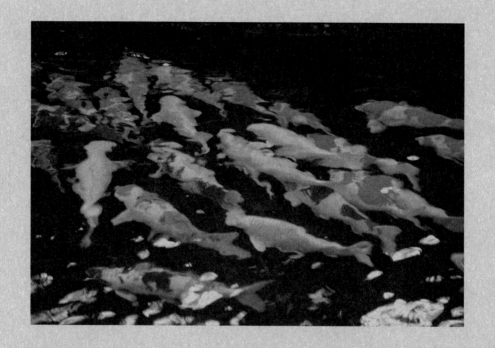

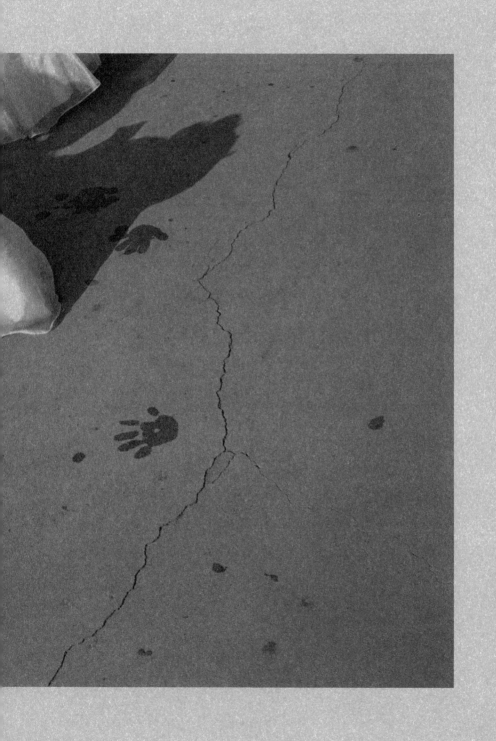

In the beginning was water

Genesis, the first Book of Moses, chronicles the biblical creation story, vividly narrating that water was already present when God began His work. Other creation myths also make it clear that water was there from the very start. Either water was an eternal, uncreated source as in the Bible, or the beginning of the world and human history was thought of as a deluge (in Babylon and South America, for instance), or life began on a small island surrounded by primeval ocean (according to some native North American nations such as the Miami).

Water is not merely a useful fluid but the basic prerequisite for all of life, the source of all being. It has the quality of being, above all else, a mirror of divinity. The Maori of New Zealand explicitly differentiate between divine water, *waiora*, and ordinary water, *wai maori*, human water, the kind used for drinking, cooking, and washing. *Waiora* is the water from certain springs known only to a select few; it is also fog, rain, and dew. *Waiora* is the medium for birth and death, and joins with tears shed in joy, grief, or farewell. Only this water is considered spiritually healing and suitable for birth and burial rituals.

According to Finnish custom, the steam rising from a glowing sauna oven, *löyly*, is a remedy for body and soul at the same time. Traditionally, heating a sauna was a solitary ritual carried out by a person whose seclusion and meditative silence were not to be disturbed. It was said that the *löyly* born from the water poured over a fire built a bridge between Earth and Heaven, body and soul, this world and the hereafter.

An intimate, even spiritual relationship to water can also thrive in a modern industrial society, as is evident in Japan. The Japanese countryside is dotted with holy places called *sei-chi*—caves, rivers, hot springs, mountains, hills, and waterfalls. Shintoism, Japan's traditional religion, sees divine spirits, *kami*, living in each one of these localities. Decorative shrines and temples are dedicated to them. Water plays a central role in Shintoism, with waterfalls being the most beautiful and touching places of power. Such places and the local divinities associated with them are honored at ceremonial rituals and annual festivals, constant reminders that nature, spirit, and awareness form an inseparable unit. The enchanting impression of holy sites, *sei-chi*, is intensified by the certainty that they served people for hundreds of years as places for divine contemplation.

Everywhere in the world, water is valued as a cleansing, transforming, and renewing element. The eternal flow of springs and rivers is a symbol of life and hope, of healing, purification, and averting misfortune. In ancient natural religions, water helped drive out from body and soul the bad spirits and demons that caused sickness. Immersion in water, an act freeing us of sins and allowing forgiveness, survives today not only in the practice of Christian baptism. Cultic and ritual ablutions and baths are found in all religions. It is common, for instance, for Orthodox Jews to perform ritual ablutions in a *mikvah*. If someone has become unclean, *tame*, perhaps by touching a dead body, he or she can again become entirely clean, *tahore*, only through complete immersion in water. In devout households, new dishes are symbolically cleaned before they are used for the first time.

Even though the hygienic side-effect of cultic ablutions is desirable, it is the spiritual aspect that is usually in the foreground. The original Christian baptism administered by John the Baptist in the Jordan River was meant to instill a mental and spiritual transformation, a turning back, a change of direction towards God. In this way, baptism even today still symbolizes the acceptance of a person into the religious community of Christians. Depending on a person's religious persuasion, baptismal contact with water ranges from a symbolic moistening of the forehead to complete immersion in a river.

In Islam, the use of running water precedes every visit to the mosque, every prayer, and every reading of the Qur'an. Most mosques have their own wells, their waters spiritually setting apart the five daily prayers from everyday life. During the minor ablution, hands, forearms, face, and feet must be washed, and one fourth of the head wetted. The major ablution is done only in an Islamic bathhouse, *hamam*. Not even the tiniest patch on the body is allowed to stay dry.

Rivers are at the center of religious experience in India, where a diversity of water rituals has been preserved to the present day. Believers and those seeking enlightenment make pilgrimages to the holy waters of the Narmada or Ganges to find redemption and healing. Proximity alone, the very presence of the river, bestows consolation and relief to believers.

In Hindu death rituals, members of the family of a person who has died bathe on odd days during the initial period of mourning, ideally in a river or a temple pond. These baths are meant to make it easier for the bereaved to break away emotionally from the deceased person, and to restore the inner balance upset by the death. Last of all, the ashes of the deceased are surrendered to the holy river.

Human beings have held water from springs and rivers in high regard since primeval times, putting water at the center of religious and cultic life. Water transforms, cleanses, and inspires, and it enables transition and metamorphosis. Waterfalls and springs are holy places endowed with power—brooks, streams, and rivers carry the desires of mankind into the world. Their flowing waters are a constant reminder of change and transience.

Human respect for water had its source not just in religious feelings but equally in experience. Springs could cease to flow, rains could fail to arrive, rivers could destroy valleys and wipe out whole villages. Water was a blessing and a danger, being a giver of life and a deadly force at the same time. No one could escape the power of water.

WHEN THE WORLD LOST ITS MAGIC

An infrequent visitor to Earth would be astonished to see that water, once so respected, merely plays an inferior role in the daily life of modern industrial society. The common perception of water today is of a 30-centimeter stream of water flowing from a faucet and disappearing down the drain a few seconds later. Water has no origin, hardly any-one knows which river or well it came from, or which soil it once rained down on. Water has become an unknown, an anonymous mass-produced commodity like so many others.

Central water supply and sewers are achievements that maintain a certain quality of life and comfort in cities, but these very achievements also have the effect of making water in pipes invisible. As a result, the consequences of withdrawing water from nature are as imperceptible to us as is the fate of dirty water after it has left our sight in household drains. Water is "domesticated" in the urban environment—streams in cities are captured in underground pipes, rivers are straightened out and embanked, any feeling of connectedness to water gets lost more and more. With our image of water characterized by puddles and ponds in city parks, and with water completely controlled and readily available, we gain the impression that modern-day humans no longer depend on it.

THE ATTRACTION OF WATER

At first glance, the modern world has removed all religious and spiritual notions from the perception of water. Dealing with water is a soberly practical endeavor, focusing entirely on usage. Only when we look more closely do we realize that our experience of water is much more faceted than we realize during daily life. Water in nature has the power to change our moods, fill us with emotions, elicit ideas, and let us connect to our innermost selves.

Water evokes emotions that vary from individual to individual, depending on where we come from, how old we are, and what mood we're in. The way water is embedded in the landscape shapes our feelings. A spring murmuring quietly from mossy ground is peaceful, young, pure, and fresh; a brook meandering through meadow valleys babbles as invitingly as a spring morning. In contrast, ragingly wild streams plummeting over blocks of stone in the mountains—so loud we have to shout to make ourselves heard—arouse our respect, deeply stir our inner selves, and can elicit courage and passion.

In October 1779, the German poet Johann Wolfgang von Goethe walked to the Staubbach waterfalls in the Lauterbrunnen valley in the Swiss Alps. It is still possible today to reconstruct what he experienced there. The Staubbach stream is a powerful torrent that pours over the rocky western flank of the valley and drops freely through the air for nearly 300 meters, on its way dispersing into misty spray. The stream touches lightly on a ledge, changes its course but slightly, and then continues to flow swiftly down towards the valley. We know from Goethe's letters that he chose to walk there on a sunny day. "We saw the Staubbach for the first time on a fine day and the blue sky shone through." It was late in the morning when his party arrived at the Staubbach, a time when spray from these free-falling glacial waters turns into constantly changing veils of rainbows. Fine drops of water swirl in updrafts so that the waterfall seems to dissolve in an airy whirlwind that rises and falls.

This sight must have deeply moved Goethe, not so much because it was a great natural spectacle, but because of the impression, mood, and associations the Staubbach waterfall evoked in him. In the days that followed, his impressions were condensed into words that are only superficially a description of the waterfall. In truth, they are a metaphor for human life. Goethe called his poem *Song of Spirits on the Waters*. His metaphor for transience and change, for human fate, is interpreted as a symbol for the cycle of death and rebirth, as one of the first expressions in western philosophy of the idea of reincarnation.

Other writers too were moved and inspired by the unique power of water. Thomas Mann in 1925 noted: "For my person, I gladly profess that the contemplation of water in all kinds of manifestations and forms signifies by far the most immediate and insistent way of enjoying nature. Certainly the real state of contemplation, true oblivion, the real dissolution of one's own limited existence into the universal is granted to me only in this experience." (*A Man and His Dog*, 1925).

In *Siddhartha*, Hermann Hesse developed an entire philosophy from his experience with water. "Isn't it this that you mean: that the river is everywhere at the same time, at the source and at the mouth, at the waterfall, at the ferry, at the rapids, in the ocean, in the mountains, everywhere at the same time, and that for the river there's only the present, not a shadow of the past, not a shadow of the future?"

It is no different for scientists when they research the properties of water as a liquid. It is difficult to obtain certainties, nothing can be grasped, everything is in constant change. Theodor Schwenk, founder of the Institute of Flow Sciences in Herrischried in the Black Forest in southern Germany, believes there is no purpose in resisting this basic quality of liquid water. "The world of water is one of movement, of becoming and fading away, of processes." Perhaps this is the very thing that has always fascinated people about water—that it makes the experience of transience immediate. "You cannot step into the same river twice," said Greek philosopher Heraclitus long ago. It should be noted that you, the bather, have also changed.

The animating, meditative effect of water has been known to humans from the very beginning. Feelings are aroused, ideas start flowing, inter-connections between all things are revealed. One of water's fundamental qualities is that it opens up paths to make contact with spiritual spheres of existence. No differently from our nature-believing ancestors thou-sands of years ago, we are inwardly moved by water. Today's approach towards water however takes on other forms, for instance, in a kind of "water renaissance." Contemporary architecture for spas and resorts in Europe in recent years has celebrated water in highly aesthetic buildings, reveling in its sound, movement, and color, and creating mod-ern water cathedrals in which reverence is combined with a sense of well-being. People are keen to come closer again to water, do something good for themselves and renew their reserves of mental and physical strength. The market for wellness is expanding as a result. Water, whether icy or warm, splashing or rushing, spurting, churning, or steaming, is *the* central medium of this modern health movement.

More and more people feel attracted to water for spiritual reasons, too. They seek more intimate contact, seeing water as a source of inspiration and healing independent of religion, sometimes even as a living being with its own emotions. Water in landscapes, in streams, lakes, and springs, is connected with elemental creatures that protect water and give it a special quality. As in fairy tales, these creatures are called nixes, mermaids, or sea goddesses.

LEARNING FROM WATER

For many people, water has retained to the present day its eminent significance as an element of purity, change, transition, and reform. It is crucial to realize that this importance must find expression in human activity. There are several indications that humanity is changing its attitude towards Earth and its central organ, water. Although commercial and political activities are shaped primarily by straightforward business interests and rational knowledge, another attitude is asserting itself more and more. This is the intuitive rejection of certain technologies, a sense that water is not a trade commodity, the conviction that rivers and lakes must remain unpolluted even if they are not needed as reservoirs for drinking water, the feeling that not every river must be kept in check behind a dam wall just because it has the potential to supply hydro-electric power.

In bonding with water, people develop greater responsibility towards it. Water in western societies is beginning to regain its value beyond use-fulness and commercial exploitation, just as certain waterfalls in Japan are holy, as are dew and rain for Maori, and certain rivers for Hindus. It appears that humanity, after a century of technically optimized exploitation, has now been awakened to realizing how many formerly wonderful, living rivers have been severely damaged, and how much the existence of everlasting water is endangered—and with it, the survival of mankind.

Water is sacred; it is a source of ideas and an oracle. The intellectual insights furnished by books, laboratories, and computer simulations are joined by our direct experience of nature and the inspiration people feel in the presence of water. We can learn to understand the language of water again, the murmur of streams, the thunder of waterfalls, the patter of rain, the roar of surf, but as well, the message heard from water when we're gargling, cleaning, cooking, drinking, or swimming. And take to heart the import of what is being confided to us all—that we come from water, that in this world we are part of an entirety.

In the beginning was, and always is, water.

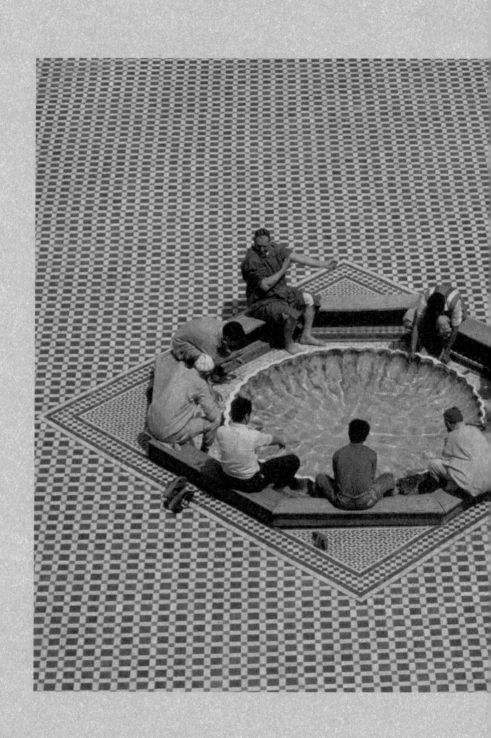

p. 409 Abbas/Magnum Photos

p. 410 Thomas Kern, USA, LA

p. 412 Claude Monet:
"Water Lillies," Museum of Modern Art, New York City, 2005.
Thomas Hoepker/Magnum Photos

p. 420 Ukraine. Odessa.
Ian Berry/Magnum Photos

p. 422 Mausoleum of Sidi Abdelaziz, Tabaa, Marrakesh, Morocco.
Bruno Barbey/Magnum Photos

WATER AND POWER

Doing business with water

Water is cheap, but supplying it to people isn't. Accessing drinking water, constructing and maintaining a mains network, and disposing of dirty wastewater are among the most expensive municipal services.

It is the emerging and developing nations that bear the brunt of this problem—their water supply and sanitation systems are still unsatisfactory in many places. Their exploding populations put beyond all thought the idea of meeting western standards in water supply and sanitation. These countries would be pleased just to achieve the United Nations Millennium Development Goals set forth in September 2000—to halve by 2015 the number of people who don't have access to safe drinking water and basic sanitation.

For more than two decades, the partial or complete privatization of municipal waterworks has been propagated as an economic solution for better coping with the expensive public responsibility of supplying drinking water. Indeed, starting in the 1980s, development institutions such as the World Bank and the International Monetary Fund granted loans under conditions that forced many Third World countries to liberalize and privatize the water sector.

We can see now that this hasn't made it easier for developing and emerging countries to overcome their enormous financing limitations. Many privatization projects have failed or been called off because of public pressure. These projects have often led to an extremely selective expansion of the water supply system, and have almost always driven up the price of water, making it unaffordable to a large portion of the population.

For a few years now, private businesses, public institutions, and civil society have been trying out new models of cooperation called public-private partnerships. The initial upshot is that these cooperative efforts are most successful when government institutions, dedicated groups in the affected populations, and local private businesses get together to work on small-scale and manageable projects to find solutions tailored specifically to local needs.

p. 428 Datah, Bali, Indonesia, 2004. Ardiles Rante/Keystone

Privatization–the great disappointment

Hardly any city in an emerging or developing country can expand its drinking water and wastewater-disposal system as quickly as the population grows in its slums. It would be hopelessly overburdened by the job, even if it limited itself to meeting the modest standards set at international development, health, and water conferences. These standards, defining the absolute minimum sufficient for dignified life, prescribe that everyone should have access to a source of drinking water within a distance of one kilometer, with at least 20 liters of water available per day.

The action plan devised at the 1977 United Nations Water Conference in Mar del Plata stated that safe drinking water and adequate sanitation should be available to all by 1990. In 1990, this target was pushed forward to 2000. In September 2000, governments at the UN Millennium Summit agreed that the number of people without sustainable access to safe drinking water should be at least halved by 2015. At the 2002 Johannesburg Summit for Sustainable Development, governments decided that the number of people without access to basic sanitation should also be halved by 2015. But experts fear it is unlikely that even these goals will be met.

This has many reasons. The water crisis is one of the most serious and complex problems humanity has ever had to manage. It is global in scope, affecting every country and continent to varying degrees, but it manifests itself in millions of local and regional situations that widely differ from each other and therefore require individual solutions. Furthermore, the water crisis is inextricably interwoven with three other fundamental challenges facing humanity today: hunger, health, and poverty. If these problems are to be solved at all, we will need to make a collective effort to do so.

Technical complexity and the lack of funds are not the only obstacles— experts estimate that it will take 100 to 130 billion dollars over the next ten years to meet the millennium goals. This sum is less than a tenth of annual military budgets. The UN Millennium Project's working group on water and sanitation has established that indeed the lack of political will is one of the biggest obstacles that keeps great deeds from following great words. This holds true not only for politicians in affluent industrialized nations, but often enough for the governments of the most affected developing countries too. It is clear why this is so. Where finances are most desperately needed, the principles of the market economy are usually not in force. The inhabitants of affected cities simply can't afford to pay the water rates which in turn would back the huge investments needed to establish a water-supply infrastructure.

How can these financial problems be solved as long as rich countries lack the will to substantially increase development funding? Policymakers, scientists, economists, and the societies affected have wrangled over this issue for thirty years and come up with completely conflicting arguments. Could the private sector be enticed to invest in areas where profit expectations are as uncertain as they are, even though private participation in many of these projects would make sense? Can political and economic structures be institutionalized to overcome these barriers?

Those international institutions looking after economic development and combating poverty in poorer countries have responded to these questions with varying strategies in the past three decades. In line with the development-policy approach prevalent in the 1970s, as pursued by the "North-South" or Brandt Commission, they initially tended to promote the construction of major dams and irrigation systems rather than securing drinking water for people. The problem was to be tackled at its roots—increased crop yields and industrialization were thought to be the keys out of the poverty trap. Growing prosperity would enable developing countries to solve their other problems themselves, allowing them to set up health, social welfare, and educational systems, and create the public infrastructures needed for water, energy, transportation, and communication.

This approach also fitted with the interests of major donor countries. It secured profitable contracts for the technology and construction businesses of industrialized nations, and was also supposed to make sure that a large part of aid money flowed back to the donor countries.

But in reality, the huge loans granted to many developing countries created a debt crisis which remains unresolved to this day. Prosperity never materialized for the majority of populations. In many places there wasn't enough money left to maintain newly built facilities, much less repair older ones. Other assumptions made by predominantly European and American planners turned out to be unrealistic as well, for instance the idea that plans developed for wealthy industrialized nations could be simply copied and transferred to poor developing nations and other cultures.

The failure of this approach, and the growing poverty of more and more people in highly populated and poor countries, prompted big development institutions like the World Bank and the International Monetary Fund to fundamentally change their strategies in the 1980s. In full agreement with powerful industrialized nations that naturally wanted their own interests as big donor countries protected, they promoted global liberalization and the deregulation of markets to encourage foreign businesses and investors to take a share in former state monopolies and public utilities. Global competition between private contractors would make sure that efficient solutions with the best value for money would be realized, that the most advanced and economic technologies would be applied, and that nepotism, corruption, and power struggles among the elites and bureaucracies of these countries would be reduced. Finally, market-driven pricing would not only ensure the expected return for private investors, but also, so they hoped, motivate consumers to use precious water economically.

These new privatization and liberalization concepts were not designed with the situation in developing countries in mind. Conceived in the think tanks of highly developed nations and meant to be applied globally, they were first implemented in the industrialized nations themselves, most radically of all in Britain in the 1980s. Margaret Thatcher's conservative government privatized not only the railways and the energy industry, but also all of the country's public water utilities. After the collapse of the eastern-bloc countries, neoliberal economics rapidly became the dominant global doctrine. To relieve strained national budgets, many countries privatized state enterprises and municipal infrastructures for transportation, communication, energy and water, garbage removal, and other public services.

The World Bank, the International Monetary Fund, and regional development banks frequently made loans to emerging and developing nations dependent on certain conditions such as privatizing local water markets and opening up to international competition. Again, this was much to the benefit of industrialized nations—liberalization in developing countries was seen as a way to open the door for western businesses to large markets that had previously been closed to them.

One of the prime examples until recently of this strategy of incorporating the private sector was the water-supply system in Manila, capital of the Philippines. In the mid-1990s, one-third of the city's 12 million inhabitants still did not have running water. The existing mains network was antiquated and leaky in many places. There was hardly a sewer system to speak of, and the public water authority, the Metropolitan Waterworks and Sewerage System (MWSS), was hopelessly in debt.

Based on recommendations by the International Finance Corporation, a World Bank subsidiary, Manila's drinking-water supply system was divided into two sectors. The eastern part of the city was put under the management of Manila Water, a company owned by a local investment group jointly controlled by the Ayala dynasty, the American construction company Bechtel, its subsidiary United Utilities, and the Japanese Mitsubishi corporation. The western part was taken over by Maynilad Water Services, a company controlled by Suez/Ondeo, the world's second-largest water company, and Benpres Holdings, also owned by a powerful local family clan. The former public water authority, MWSS, which had been the owner of the water systems, changed into a regulatory body.

Thanks to both companies' generous bids, they were awarded concessions for twenty-five years. They promised to supply all of Manila's inhabitants with running water, reduce the price of water by 44 percent, and at least halve water loss, all within ten years. They also planned to invest 7.5 billion dollars to build a comprehensive sewer system by 2021. Maynilad even pledged its concession fees towards covering 90 percent of the debts accumulated by the public water authority MWSS. In turn, the authority agreed to supply both companies with as much water as they needed at all times at no cost.

Only five years later, the experiment turned out to be a complete fiasco for all concerned. Both companies had almost met their goal of supplying running water to some two million new customers by 2002, if their generously interpreted figures are to be believed—Maynilad reckoned that each new connection delivered water to a household with nine persons. But the question was unanswered, even largely ignored, of how to supply clean water to those several hundred thousand inhabitants too poor to pay expensive connection fees.

Maynilad fell far short of all other goals. Water loss was not halved, indeed it increased—at the expense of customers and taxpayers. Maynilad passed the cost of growing water loss on to consumers without keeping its promise to repair the network. The water authority was forced to tap additional water sources at ever-increasing expense—this water then seeped away somewhere in the privatized mains network. Contrary to all declared intentions to keep water rates at their original level for at least ten years, rates were already considerably higher after six years than before privatization—and four times higher than originally agreed.

Early in 2001, the regulatory authority only partially approved a request from private operators to increase rates again. In response, Maynilad in March 2001 stopped paying the agreed concession fees intended to pay off MWSS' old debts. By mid-2004, outstanding payments from Maynilad amounted to 180 million dollars. Maynilad terminated its concession contract in December 2002. The reasons given were the Asian Crisis, the repeated devaluation of the Philippine currency, insufficient rate hikes, and the effects of El Niño. More importantly, Maynilad claimed that the regulatory authority MWSS had breached its concession agreement. In turn, Maynilad demanded 300 million dollars in compensation.

Whether the case is settled within or outside of court, Maynilad will withdraw from Manila, and taxpayers and water customers will have to bear the brunt of the financial disaster and pay off an enormous debt. In the meantime, the city has not come any closer to having a modern and sustainable water supply system.

In 1998, the World Bank approved a loan of 25 million dollars to the city of Cochabamba in Bolivia on condition that it turn over its municipal waterworks to a private consortium. The World Bank also demanded that all investment and operating costs be financed by water rates—under no circumstances was any of the loan to be "misused" for subsidizing water supply to the poor. When the price of water jumped by 35 percent shortly after the consortium took control, tens of thousands took to the streets. The Coordinadora de Defensa del Agua y de la Vida citizens' action group verified that a majority of the inhabitants in Cochabamba's slums were spending more money for water than for any other food. A poll indicated that more than 90 percent of the population wanted the privatization of the waterworks to be cancelled.

After violent demonstrations and a general strike, the Bolivian government agreed to municipalize part of the water supply system again. Wherever some areas stayed in private hands, the government negotiated new

First recital of the EU
Water Framework
Directive, 2000

Water is not a c
like any other, b
which must be p
and treated as s

Peter Brabeck-Letmathe,
CEO of Nestlé, Vevey,
Switzerland, in: We feed
the World, Erwin Wagen-
hofer, Max Annas

Water is a foods
other foodstuffs
a commercial va

mmercial product
t rather a heritage
otected, defended
ch.

uff, and like all
t ought to have
ue.

contracts. Since then, modernization of the water-supply system in Cochabamba has been somewhat more successful. The population affected has taken an active part in the decision-making process and in construction and maintenance of the system as well.

In the past few years, many other cities in both rich and poor countries have had similarly negative experiences with the privatization of waterworks, from Atlanta, Georgia, to Jakarta in Indonesia, from the French tourist town of Grenoble to Halifax, Canada and Cartegena in Columbia, and from the Argentinean capital Buenos Aires to the eight largest cities in Mozambique. Under pressure from the International Monetary Fund and the European Union, the water supply system in Mozambique was handed over in 1999 to a consortium headed by the private French water corporation Saur. When Saur withdrew from the contract in 2002 without giving any reasons, the Portuguese state water company, Aguas de Portugal, active in other former Portuguese colonies as well, took over.

In 1999, the Berlin Senate sold a 49.9 percent share of its municipal waterworks to a private consortium made up of two international water suppliers, RWE and Veolia, for 1.58 billion euros. Administrative and financial management was transferred to a new company, Berlinwasser Holdings. In a contract kept secret from the public for years, the city guaranteed the consortium an annual interest rate of approximately 8 percent over a period of twenty-eight years as return on investment. Annual consortium investments increase total operating investment by two hundred million euros each year, thus increasing the guaranteed return by 16 million euros each year too. Until 2004, this return was financed by successive waivers of the city's share of profit, meaning that it came indirectly from the city budget. Since 2004, water rates have been hiked several times to generate the guaranteed return. Independent experts have calculated that Berlin would be much better off today if it had not partially privatized its waterworks.

Opponents of privatization argue that the involvement of private water suppliers in the water economy has failed so spectacularly in many cases, or at least not fulfilled expectations, because there is a fundamental contradiction between state institutions being committed to the public good and private enterprises being committed to the interests of a relatively small group of owners and shareholders. A private business can't make an investment if it doesn't expect profits. It's not a coincidence that only 0.2 percent of private water investments around the world are made in sub-Saharan Africa. It's also not a coincidence that the fiercest conflicts arise with private water suppliers wherever a majority of consumers can't afford to pay fees for connecting to the water mains.

Since the 1960s, water supply has been at the center of a grim debate on the advantages and disadvantages of globalization, liberalization, and privatization. This controversy has not been resolved, and it pitches rich industrial nations and their corporations, on one side, against many developing countries, unions, and a large number of civil society organizations on the other. Persisting criticism has at least led to a more detailed analysis of why so many privatization projects fail. Even though the ideological debate continues unabated, conflicting parties in many places have joined forces in pragmatic teamwork. Many different forms of cooperation between the public and private sectors, called public-private partnerships, are currently being tried out.

We still don't know if these pragmatic approaches will be a better way of solving existing problems. It probably depends mostly on how far traditional decision-makers in the water sector are willing to allow new civil-society actors to actively participate in decisions.

p. 438 Popular protests against the private water company Suez Lyonnaise des Eaux over failing water supplies and high prices, El Alto, Bolivia, 2005. Martin Alipaz/Keystone

Private water supply–who controls the market?

The enthusiasm of financial analysts around the world is still great, even though cautioning voices are starting to be heard. The water business, they predict, will be a safe and highly profitable capital investment in coming decades. In 2004, the global volume of business for private water suppliers amounted to more than 400 billion dollars. As early as 1998, the World Bank predicted a doubling of this turnover for the near future, and it even increased its forecast in 2001 to over one trillion dollars. Private water companies currently service only five percent of all connected households. Financial experts believe that in coming years, many towns, cities and large urban centers with shortfalls in their budgets will hand their public water utilities over to the private sector. The International Monetary Fund and regional development banks foster this trend by making their loans dependent on the liberalization and privatization of water markets.

Considering this rosy outlook, it is a bit surprising that fewer than a dozen international corporations have divided up practically the entire private water market among themselves. The sector is dominated by two powerful French corporations, Suez/Ondeo and Veolia Water, who together control more than two-thirds of the world market. This is hardly accidental. France was the only country in the world to have already largely privatized its water utilities in the 19th century. A hundred years later, after the collapse of the eastern bloc, when the liberalization of markets and the privatization of public services like mail delivery, railways, and power and water supply became accepted global economic doctrine, French water providers had already made an almost unbeatable start.

Although in the late 1980s many cities around the world privatized their water utilities, the number of large water providers on the global market hardly grew at all. Even powerful North American multi-utility companies who seemed predestined for the task made only limited investments in water services.

Suez/Ondeo, a multi-national group providing water to 115 million people and active in 130 countries today, was created in 1997 when Lyonnaise des Eaux and Compagnie de Suez merged. Each company had been established more than a hundred years earlier, and both were already internationally active before the merger. Suez, a finance and industry group, acquired in 1994 a substantial stake in the U.S. company United Water Resources, later taking it over completely. Of the thirty largest cities which ceded their waterworks to private operators between 1995 and 2000, twenty sold or awarded concessions to Suez-Lyonnaise or one of their predecessors. In 2002, the new group, which is active in the energy and waste-disposal sectors as well, merged all its water operations under the name of Ondeo. By pursuing an aggressive strategy of expansion, especially by buying up many companies throughout the world, Suez/Ondeo was able to rapidly expand its global position.

The company had an annual growth rate of about 25 percent, but this was at the cost of increasingly high levels of debt, hitting 29 billion dollars in 2001. This forced Suez/Ondeo to concentrate on its core business of utilities at the same time, withdrawing from other sectors such as transport and telecommunications. Instead, Suez/Ondeo further strengthened its water operations in Europe and the United States by buying US Water, a subsidiary of the Bechtel industrial group, which provides water to some 40 million Americans.

Due to massive problems in many emerging countries in the grip of financial crises, Suez/Ondeo has withdrawn in recent years from megacities like Manila, Buenos Aires, and La Paz. The company's 2003 plan for action foresaw it reducing its activities by one-third in developing and emerging nations viewed as unstable, and limiting global investment in physical assets to the bare minimum in years to come. Suez/Ondeo intends to focus its activities wherever possible on joint public-private ventures that promise more profits, for example in logistics and administration.

Veolia Water, the second-largest water company worldwide, is following a similar strategy. Veolia Water is a subdivision of the French environmental-service provider Veolia Environnement, which until 2003 was owned by the scandal-ridden conglomerate Vivendi. Before changing its name to Vivendi in 1997, the more than one-hundred-year-old Compagnie Générale des Eaux had already expanded into many other business sectors. The group consisted of about 3,000 energy, water, and waste-management companies, a chain of department stores, real estate holdings, construction and transport companies, as well as media and communications businesses.

Under the management of Jean-Marie Messier, Vivendi from 1997 focused on the two sectors that seemed most profitable–media and environmental services. Most of the businesses in other sectors were sold. Vivendi then proceeded to acquire television broadcasting stations, film and music groups like Seagram with Universal Studios, the German Babelsberg film studios and the Universal Music Group, the world's largest label in the music industry, as well as book and magazine publishers, telecommunications companies like AOL, and some of the biggest companies in interactive electronic entertainment.

Messier combined the environmental activities in energy, water, and transport into one subsidiary, Universal Environnement, of which he floated one-third on the stock market. Since this business sector had immense material assets, Messier borrowed heavily on it to reinvest in the media and telecommunications sector.

These risky financial transactions resulted in a debt burden to Vivendi of 35 billion euros, the consequences of which were felt mostly by water and energy customers. This and several bribery and corruption scandals forced Messier to resign. A bank consortium saved the conglomerate from bankruptcy. Large shares of the business were sold to consolidate it, including in 2003 the majority of Vivendi Environnement, which has since operated under a new name, Veolia Environnement, to help customers forget the negative image of its former parent company.

The four subdivisions of Veolia Environnement, Veolia Water, Onyx (waste management), Connex (transport), and Dalkia (energy), continue to be active around the world in many subsidiaries. Veolia Water currently services about 80 million people in several dozen countries. As in the case of its competitor Suez/Ondeo, most of these operations are in Europe and the United States. Only 6.5 percent of turnover is generated on other continents. To return to profitability, Veolia has been forced to sell some valuable subsidiaries in the United States, including USFilter, the market leader in water-purification technology. In the future, Veolia intends to focus on safe, financially stable markets and withdraw from some developing countries.

On the international market, the only business that can keep pace with the two French conglomerates, at least until 2007, is the German energy and water utility RWE (Rheinisch-Westfaelische Elektrizitaetswerke), together with its powerful British subsidiary Thames Water. RWE was originally a merger of several municipal energy utilities in Germany and has been largely in private hands only since 1998. It established itself as a conglomerate in the 1980s with well over a hundred subsidiaries active in the coal, gas, oil, nuclear power, and waste sectors. RWE also owned or still owns a chain of gas stations, printing-machine factories, chemical plants, construction companies, telecommunications companies, and other businesses.

In 2000, RWE became a significant global player in the water business when it bought Thames Water, the U.K.'s largest water utility, which itself owns a number of subsidiaries, mostly in Asia, Australia, and Africa. In 2003, RWE/Thames Water bought American Water Works as its most important acquisition for the future. This American business has fifty-five subsidiaries in the U.S. supplying water to more than 15 million people, and it's also broadly active in Latin America. RWE/Thames Water in 2002 was able to acquire half of China Water Company, thereby gaining a promising start in one of the largest emerging economies.

But late in 2005, RWE announced it would withdraw from the international water market. It intends to sell Thames Water and American Water Works before the end of 2007. RWE declared that the purchase of these two companies for altogether 16 billion euro had not been worth the investment. In London alone, several billion euros will have to be spent in coming years to repair and modernize the city's decrepit network of 96 thousand kilometers of water and sewer pipes. The debt-ridden parent company doesn't have this kind of money. Of the 3.5 billion euro which RWE recently invested on average each year in all its business divisions,

God provided th the pipes.

Gérard Mestrallet, CEO of the multinational corporation Suez-Ondéo, whose operations include water.

Subrata Mukherjee, Mayor of Kolkatta (Calcutta), India, in reaction to a reported letter from the central government asking the West Bengal state government to ensure privatization of water supply in urban areas.Times of India, 6 Jan 2003

Fifty percent of p city are supplied If supply gets pr ing the slum dw pay. This we car

e water but not

eople living in the
water free of cost.
atized, all includ-
lers will have to
t allow to happen.

43 percent went to its two water subsidiaries alone, although their turnovers of 4.1 billion euros made up less than 10 percent of RWE's total turnover.

Besides these three giants, there are only five or six other groups of any significance in the international water business: the French Saur Group, a subsidiary of the French construction corporation Bouygues, with subsidiaries mainly in African countries, but also in Argentina, Poland, and Vietnam; the U.K. corporations Anglian Water, Kelda Group, United Utilities, and Severn Trent, each, however, with just a few million customers; and the American Bechtel/Edison group.

In an attempt to enter the international water market, E.on, Germany's largest energy utility, bought the Gelsenwasser water company in 2000, which was already well-positioned within the country. But three years later, in the course of E.on's merger with Ruhrgas, regulatory authorities demanded that E.on resell Gelsenwasser. Rather unexpectedly, the new owners were not private interests but the cities of Bochum and Dortmund, who intend to run Gelsenwasser, which is also active in Eastern Europe, according to free market principles.

In practice, the private water market is a highly volatile and fiercely competitive sector in which huge profits and equally huge losses can be posted in back-to-back business quarters. Within a very short time, the largest groups have been reduced from reliable technology businesses with long-term prospects to mere playthings of erratic financial speculators. Most of them have been repeatedly restructured in the last ten years. These businesses have attempted to compensate for absent profits, and to minimize abruptly growing debt, with shortsighted, wild swings in strategy as well as acquisitions and sales. Huge multinational groups upon which millions of water customers depend unexpectedly experienced financial difficulties, like Vivendi did, or else they completely vanished from the market. The most eminent example was America's Enron Corporation, which vigorously entered the global water market in 1998 through its subsidiary Azurix only to withdraw three years later, the same year it declared bankruptcy with a debt burden of 13 billion dollars.

In many cities, water rates increased after privatization, while investments often lagged behind what had been contractually agreed. Wherever profits did not meet expectations, businesses retroactively attempted to renegotiate concessions to their advantage. In many cases, they took municipal authorities and regulatory bodies to court or prematurely withdrew from their contracts.

Private water utilities were in turn confronted with fierce opposition whenever they failed to meet the expectations of municipalities and their inhabitants. In cities like Manila, Johannesburg, and Buenos Aires, and in places like Puerto Rico and Mozambique, authorities accused private concessionaires of not investing in the expansion of water networks and replacing leaky pipes as they had promised. In Cochabamba, Jakarta, Atlanta, Halifax, Vancouver, and numerous European cities, people protested loudly and municipal authorities declined to renew concessions or terminated them prematurely. The Netherlands no longer wants to expose its population to such risks, and since 2004 has legally forbidden private businesses from acquiring the assets of waterworks.

Meanwhile, financial analysts' enthusiasm has somewhat cooled. Experiences to date have shown that one of the biggest challenges facing humanity—the sustainable provision of safe water—is perhaps not something that lends itself to becoming an object of short-term financial speculation. Within just a few years, a handful of global water suppliers went through surprising crises and disasters, changes in strategy, mutation, and restructuring—hardly raising the confidence of municipal authorities and especially local populations in their services. Expectations lie elsewhere—in high-quality water, long-term water accessibility, integrity, and not least, social responsibility towards underprivileged members of society.

p. 448 Gordon Brown, Chancellor of the Exchequer, Great Britain; Paul Wolfowitz, President of the World Bank; Ngozi Okonjo-Iweala, Nigeria's Minister of Finance, 2006. Manuel Balce Ceneta/Keystone

Expensive bottled water

No other part of the food and beverage sector grows as rapidly and apparently unstoppably as the bottled-water industry. Some 150 billion liters were marketed in 2004 and forecasts say this figure will go up 10 percent each year. Bottled water manufacturers expect to sell up to 265 billion liters in 2010.

Half of all bottled water is still sold in Western Europe–more than 100 liters per person each year. The Italians top the list with 190 liters, followed by the French (150 liters), and the Germans (110 liters). The market is nearly saturated in these countries, but bottled-water companies expect high growth rates in many other regions for a long time to come. The United States is the world's biggest market for bottled water, but at currently 70 liters per person each year, it is far from reaching a limit. Consumption in Eastern Europe has practically doubled since 1996, but at 20 liters per person it lags far behind expectations. This applies even more to the big markets in Asia, Latin America, and Africa, where annual per capita consumption is still only a few liters.

The hectic efforts of some major multinational companies to secure market shares show just how much growth potential the food and beverage industry sees in the sale of bottled water. Up until the 1980s, the mineral water business, drawing on cheap and very abundant resources in the United States and Western Europe, was considered rather unspectacular and was run almost exclusively by small local companies. The local character of the business evolved because it's not the contents of the bottle that incur the most costs but the transport. A liter bottle of water weighs one kilogram, not counting the weight of the bottle itself.

Today, four corporations, including the Swiss food and beverage giant Nestlé and its French competitor Danone, dominate 40 per cent of the global market. The expansion and consolidation process is continuing at full speed.

Market leaders Nestlé and Danone's predecessor BSN started to buy up bottled-water companies in 1969. Nestlé acquired 30 percent of France's Societé Generale des Eaux Minérales de Vittel, and BSN bought up Evian/Volvic. But they didn't start investing heavily in the bottled-water industry until the mid-1980s. Between 1992 and 1998, Nestlé acquired Perrier and San Pellegrino, the market leaders in France and Italy, and numerous smaller companies. Danone concentrated its investments on Asia, Eastern Europe, and Latin America where it bought up many regional and local companies. Today Danone owns Indonesia's largest bottled-water producer and two large Chinese bottled-water manufacturers as well.

Maneka Gandhi, Indian
Member of Parliament
for the Bharatiya Janata
Party (BJP), 21 Novem-
ber 2005

These compani water, put in a p imported from t and then sell us of INR 12 (EUR

s take our ground-
nch of essence
e United States
ur water at a price
22) per bottle.

Meanwhile, the two largest American soft-drink producers, Coca-Cola and Pepsico, have launched their own brands, Dasani and Aquafina, and the food group Unilever has also entered the mineral-water market on a large scale. These companies have excellent prospects since all three have extremely well-developed global distribution networks. Latecomers Vivendi and Suez/Ondeo, the world's two largest suppliers of drinking water, are having a much harder time because they lack an efficient distribution network for retail delivery.

As in other young and promising markets, the bottled-water business is highly active, in particular because the four leading corporations compete aggressively with each other in neighboring food, soft drink, and wellness markets—even changing their alliances when it suits them. Nestlé and Coca-Cola have founded a joint distribution company, Beverage Worldwide Partner, which supplies kiosks and vending machines around the world with Nestlé's cold tea and coffee beverages, and soft drinks from Coca-Cola. At the same time, the two corporations are competing against each other on the bottled-water market in North America, where Coca-Cola has joined forces with Nestlé's rival Danone to beat its own competitor Pepsi.

To consolidate their positions in this contested market and increase profits by offering additional services, bottled-water manufacturers early in the 1990s started to expand to every part of the globe their sales of water for dispensers in homes and offices. Originally limited to the United States, and known as "home and office delivery," this business now makes up about half the turnover in bottled water in emerging countries. In Western Europe, this share is only about 3 percent.

In the luxury and lifestyle sectors where high-end brands such as Perrier, San Pellegrino, Evian, Valserwasser, and exotic brands like Scottish Highland Springwater are sold at increasingly exorbitant prices, bottled-water manufacturers can indeed expect long-term growth rates. In contrast, forecasts are very uncertain for cheap, purified bottled water intended for people who have no access to running water in the outskirts and slums of cities in developing countries. Local women's initiatives and community groups in these countries have long resisted the dubious fortune of being supplied with water in a bottle.

In some cases, bottled water is a necessary supplement when tap water is insufficient or even lacking in areas affected by severe water scarcity. But in other places, bottling plants are the cause of increasing water scarcity. In Plachimada in the Indian state of Kerala, a Coca-Cola factory extracts 1.5 million liters of water from the ground every day. In the surrounding area, the water table has dropped by some 100 meters and many wells used by the population for drinking water have run dry. "The local adivasi women had to travel about five kilometers to fetch drinkable water," reported the Mathrubhumi daily newspaper on March 10, 2003. "During the time this took them, soft drinks would come out of the plant by the truck-load." (→ pp. 392)

It's for very good reason that poor slum dwellers in megacities protest against having to buy bottled water when they should be supplied with the running water that has been promised to them—especially when drinking water flows freely from faucets in prosperous neighborhoods not so far away and is fifty to a hundred times cheaper.

p. 456 Baghdad, Iraq, 2003. David Leeson/Keystone

On the road
to new partnerships

A March 2002 discussion paper on the World Bank's strategy in the water sector stated that: "Under current conditions the private sector will play only a marginal role in financing water infrastructure." This skeptical assessment contradicted the doctrine the World Bank had applied until then. Since the early 1990s, the World Bank had granted loans to developing countries on the condition that they privatize their water management, believing that only the private sector would be able to make the large investments needed.

In reality, the commitment of international water corporations falls off more every year in developing and some large emerging countries. The Suez/Ondeo water group plans to reduce its investments in the Third World by one-third. Veolia plans to completely withdraw from some countries (→ pp. 441).

Water corporations have not seen expected returns on their investments. In more and more cities, regulatory authorities have refused to approve the hikes in water rates that companies request. Wherever authorities have given in to pressure from water companies, protests from the local population have often forced a reversal of rate increases, or even in some cases the premature termination of concessions.

As early as 1996, a World Bank report on privatization in six European, Latin American, and African countries criticized the policies of private water companies as ruthless and unsuitable to the circumstances. At the 3rd World Water Forum in Kyoto in March 2003, representatives of the World Bank and water corporations both confirmed that they had failed to assess the situation correctly. Nemat Safik, director of the World Bank, admitted that: "We were too optimistic concerning the willingness to invest in these countries. Despite far-reaching reforms, many countries do not find investors." Olivio Dutra, a Brazilian government official responsible for urban development, confirmed that: "Privatization has not resolved the water problems for most of the population." Richard Aylar, head of Thames Water, scoffed: "Somebody is overselling the idea of privatizing water as a way to reduce poverty."

The hope that all problems of water supply could be solved by involving private enterprises has proven false. Corporations often had unrealistic expectations of profits and were ignorant of local conditions. In turn, the desire of municipalities to hand over a difficult problem to the private sector often led them to sign agreements in which both sides made promises neither could keep.

In recent years, private water companies have increasingly attempted to shift their cooperation with municipalities to more manageable and less risky projects. The range covers everything from the relinquishment of management, logistics, and controlling to the sale of parts of water-works, from narrowly defined construction and operating commissions to lease contracts and operational concessions with clear profit guar-antees and cost-sharing agreements obliging municipalities to bear any eventual risks which may occur, like currency exchange losses.

A new idea called public-private partnership promises a better way to handle the cooperation between public and private sectors. Although widely discussed, the content and extent of these partnerships still need better definition. The term was coined partly in reaction to massive worldwide protests against neoliberal global privatization and liberaliza-tion doctrines. Its conciliatory message is that opponents become partners. Although the idea initially functioned as a kind of reassuring rhetoric at international conferences, it has actually led to a range of sustainable and successful projects on the local level.

The novel aspect of this approach is that, for the first time, civil society organizations have been accepted as third partners with state and private sectors. Those who are directly affected–local initiatives, women's groups, farmers' cooperatives, non-formal neighborhood groups, environ-mentalists, church and charity organizations, and international aid organi-zations–are heard and included in decision-making processes. This opens the door for many different, new, and sometimes unconventional solutions that are much better tailored to local and regional conditions. The specific needs of those concerned, and their social and cultural backgrounds, are taken into account. Ecological issues are voiced. Many projects can be built with cheap materials available locally. Native work-ers can employ traditional, familiar construction methods, and local craftsmen and small businesses can contribute their experience and skills. Those affected, including women as well as uneducated and unem-ployed people, can all be included and find temporary, perhaps even permanent sources of income. In this way, construction costs are reduced and the local population's commitment often ensures that users feel responsible for maintaining and looking after their facilities once they have been built.

We can't foresee how this new partnership approach will work out in the long run. Will it be applicable beyond the problem areas of the Third World where previous solutions have so broadly failed? Will previous decision-makers, whether in the public or the private sector, be willing to allow civil society organizations to genuinely participate on a broad level? Who will make final decisions when it comes to a conflict of interests between the needs of those concerned and the commercial expectations of the businesses involved? Representatives of Aquafed, a newly founded association of private water companies, are already warning against allowing civil society too much influence. According to spokesman Jack Moss, they fear that: "They are used by small unrepresentative minorities to obstruct democratic decision-making," with "the danger of disenfran-chising duly empowered democratic representatives."

" I am convinced water in genera a private busine service instead.

Evo Morales, President of Bolivia, when asked about the decision to oblige El Alto water utility Aguas del Illimani (Aisa) to leave the country.

hat potable water–
-cannot be
s, but a public

A number of international agreements currently being negotiated within the World Trade Organization (WTO) are even more crucial to the future of the water-supply sector and especially the role of private enterprises. They could fundamentally change the political conditions for liberalizing water markets. Decisive is the General Agreement on Trade in Services (GATS), a service agreement now being renegotiated that first went into effect with the founding of the WTO in 1995. It obliges its 148 signatory states to open their service markets to foreign providers and remove any possible trade barriers. Domestic businesses should not have an advantage over foreign investors. Added to that, any advantages granted by one WTO member to a country should be accorded to all other WTO members. Exceptions to these liberalization regulations are some regional free-trade zones like the European Union and the North American Free Trade Agreement (NAFTA) countries—precisely those markets that belong to the richest industrial nations, the staunchest supporters of liberalization.

Critics argue that GATS is a one-way street leading to the complete privatization of the service sector, including those areas previously considered the responsibility of the state. They fear that private control of this sector could not be reversed even if it completely failed. Individual states' present scope for action would be very limited. The WTO could object to and even overrule all national laws or regulations, all privatization restrictions, environmental measures, and quality standards if these limit free trade "more than necessary" or discriminate against foreign enterprises. It is unclear what "more than necessary" actually means. Critics are not alone in suspecting that this vague phrasing will give rise to many legal disputes.

Since GATS also covers investment in foreign countries, the WTO can stop even the most needed public services from being subsidized, such as supplying drinking water in slums, should foreign investors feel this puts them at a disadvantage. GATS critics fear that the treaty will force developing countries to open their service markets to competition against their will as well as lower environmental and health standards to make investment attractive to international corporations.

In November 2001, the WTO ministerial conference in Doha agreed to renegotiate Article XIX in GATS, which covers services. These negotiations were to be conducted "on the basis of progressive liberalization" and WTO members would be obliged "to enter into successive rounds of negotiations to progressively liberalize trade in services." The intention was to deal with the service sector more consistently than before and include basic public services such as health, education, and most notably water and sanitation systems. Preliminary negotiations were to be completed by December 2003 so that the new GATS treaty could go into effect in 2005. But with the collapse of the WTO ministerial conference in September 2003, in which GATS played only a minor role, the entire negotiation process was delayed.

The new GATS treaty is supposed to be negotiated behind closed doors in an extremely complicated procedure. During a first phase, individual countries are supposed to make reciprocal offers and demands regarding the opening of their markets. These would be settled against each other in continuing negotiations. According to the principle of fairness, whenever two countries agreed to open their markets to each other, their mutually-agreed arrangements would apply to all other WTO members as well.

Critics consider this procedure to be rather shady horse-trading that discriminates against developing countries that often have no choice but to give way to the demands of industrialized nations. Indeed, it has since come out that the European Union, acting in the interests of its big water corporations, is calling for the liberalization of water markets in seventy-two countries, but is not willing to generally open its own. The EU is arguing that it has already made enough concessions to developing nations regarding agricultural reform.

The new GATS treaty would endanger many projects in developing and emerging countries trying to solve their problems without the help of private businesses. This is the case in Honduras and Tunisia, where government policies on water supply have found praise even with the World Bank. Or in Bolivia, where the government founded SAGUAPC, a nonprofit cooperative that successfully runs the waterworks in the city of Santa Cruz. Or in Nicaragua, where the grassroots organization Movimiento Comunal Nicaragüense runs a number of well-functioning local waterworks, especially in rural areas. Or in places like Puerto Rico, Argentina, the Philippines, and South Africa, where experiments with privatization have failed and cities have returned the water supply system to the public sector. The fear is that in all these countries, municipalities could be forced to enter risky and incalculable privatization, even when public utilities work to the satisfaction of the population.

    p. 464 Ahaus, Germany. Wolfgang Rattay/Reuters

Water–a hot political issue

Water knows no boundaries–a fact that holds enormous potential for conflict. Wherever several countries have to share the water from lakes, rivers, or even whole river systems, they must agree on how they intend to use and safeguard this water in common. Their agreements can be cooperatively and peacefully negotiated through diplomacy, or they can be coerced through economic sanctions and military might.

There are 202 countries in the world, and 145 of them share one or more rivers with other countries. Rivers that cross boundaries hold 60 percent of the world's river waters. Forty percent of the world's population lives in the basin of an international river, and twelve countries are almost completely dependent on water from rivers whose sources are in other countries. If catchment areas are included, 261 rivers cross several countries. Nineteen rivers are shared by more than five countries and the Danube is even shared by seventeen.

The more that individual countries depend on these waters, the more the potential for conflict goes up. Finding cooperative solutions can become even more difficult when problems with water supply are aggravated by political, social, and ethnic conflicts, and when countries invoke ancestral rights, or simply make use of their superior economic or military power.

A study commissioned by the United Nations reported that 1,831 conflicts over water occurred between 1949 and

1998. Of these, a mere 37 involved significant force of arms. But 93 conflicts involved, in no uncertain terms, the threat of military, economic or diplomatic sanctions.

"Many of the wars this century were about oil, but those of the next century will be over water." This forecast, made in 1995 by Ismail Serageldin, former vice president of the World Bank, may be exaggerated, but there is no doubt that conflicts over water will significantly increase over the coming decades. Freshwater is scarce and getting even scarcer. Much will depend on the availability of water resources to individual countries, particularly in arid regions with rapidly increasing populations. A country that doesn't have enough water can't feed its people and hardly has a chance to develop economically. Its very survival is essentially endangered, as in the case of the future state of Palestine.

Since the 1950s, conflicts caused by water pollution rather than water scarcity have been on the rise. As long ago as 1987, the United Nations World Commission on Environment and Development published the Brundtland Report, which established that environmental crises had become a significant source of political unrest and international tension. But even today we still lack the international water legislation that could regulate the rights and obligations of water users and their relations to each other.

p. 468 Jiamusi, Heilongjian, northern China, 2005. Ng Han Guan/AP Photo
p. 470 The Xiaolangdi Dam on the Yellow River, China, 2005. Zhang Xiaoli/Keystone

Water is power

No commander has ever described the goal of a water war more concisely. "For yet seven days," God said to his faithful follower Noah, "and I will cause it to rain upon the earth forty days and forty nights; and every living substance that I have made will I destroy from off the face of the earth." According to Genesis, almost all the world's population at that time did in fact die. The Epic of Gilgamesh tells of a similar water war waged against humanity by the Babylonian god Ea.

Compared to such godly punishments, mankind's water wars seem to be mere skirmishes. But through the ages, water has played an important role as an efficient and versatile weapon in wars. Before the invention of poison gas and the atomic bomb, water was the only weapon which could destroy thousands or tens of thousands of people at one time. Merely threatening to destroy a city's drinking-water supply or a country's irrigation system was sometimes enough to make the enemy surrender.

It is no coincidence that the first historically recorded water wars took place in dry regions like the Euphrates and Tigris Basin. Highly developed and centralized irrigation systems were the backbone of the Sumerian and Assyrian civilizations. From a national security point of view, they were also existential risk factors. Whoever controlled the water supply had control over a whole realm's vital line of survival.

But even in regions of higher rainfall, water scarcity was still thought of as a weapon. One relatively modern idea dates back to 1503 when the Florentine strategist Machiavelli, with the imaginative help of Leonardo da Vinci, devised a plan to divert the course of the Arno River around the city of Pisa with the intention of depriving the city of water and forcing it to surrender. Ambitious as it was, this plan failed.

The tradition of threatening the enemy with as much damage to the civilian population as possible continues to the present day. City water supply systems were targeted and dams and canals were bombed to cripple and demoralize the enemy during the Second World War, the Vietnam War, the war against Serbia, the Iran-Iraq war, and in Afghanistan, Chechnya, and both Gulf wars, with the intention of forcing capitulation. Warring parties have never seemed particularly perturbed about routinely violating humanitarian international law. The Fourth Geneva Convention on the Protection of Civilian Persons in Time of War, signed on August 12, 1949, obliges signatory states to protect civilians "in all circumstances." Nevertheless, water wars are always directed first and foremost against civilian populations.

As devastating as the effect of using water as a weapon in warfare may be in individual cases, destruction is limited to a certain place and time. All damage caused by water, regardless of how great, can be repaired within a few years. In contrast, wars waged *over* water, and not *with* water as a weapon, take on an entirely different dimension. They destroy far more than a city's water supply, a dam, or irrigation canals. In the long term, they threaten the conditions of life for people in a whole region, their chances for social and economic development, indeed the very means of existence of entire countries and their populations. If one country conquers the larger portion of the water resources it shares with another, it controls the other state's most important means of subsistence.

Thousands of years after the Babylonian wars, countries in the Euphrates and Tigris Basin are still bitterly fighting over water. Both Syria and Iraq, countries with rapidly growing populations and ambitious development goals, are widely dependent on water from the two largest rivers in the region. Syria gets 90 percent of its surface water from the Euphrates, and Iraq depends upon both rivers for 60 percent of its water. The sources of both rivers are in Turkey.

Since the mid-1960s, Turkey has had its own ambitious plans for the waters of the Euphrates and Tigris. The Southeastern Anatolia Development project foresees the construction of altogether twenty-two dams and nineteen large hydroelectric power stations along both rivers on Turkish territory, generating over 26 billion kilowatt hours and increasing Turkey's current electricity output by more than 25 percent. It is also supposed to turn 1.6 million hectares of unusable land into fertile farmland. This is equal to 80 percent of Iraq's total cultivated area. Several large tourist destinations will receive generous supplies of water and one million new jobs will be created. But the project is controversial, even in Turkey. More than 100,000 people, mostly Kurds, will have to be resettled. The planned Ilisu Dam will submerge the ancient town of Hasankeyf, one of humanity's oldest archeological sites (→ p. 507).

Even though only parts of the project have been completed, Syria and Iraq are already feeling its consequences. In drought years, the level of the Euphrates greatly drops, sometimes leaving Syria's Assad Dam only two-thirds full, and forcing several power plants important for electricity supplies to Syria and Iraq to substantially reduce their output. Should the project be completed, the amount of water remaining for Syria and Iraq will be reduced by up to 60 percent.

The consequences are already partially irreversible. Some 97 percent of the fertile estuaries in the Persian Gulf have turned into barren salty deserts due to the lack of freshwater. The United Nations Environment Program describes it as an ecological disaster comparable to that of the Aral Sea. Even when water still flows, it is so contaminated by agricultural pesticides and industrial and household wastewater that it can't be used for drinking or irrigation.

Tensions have flared between the three countries since the mid-1970s. At that time, when Syria inaugurated its al-Thawra Dam (known today as the Assad Dam), a conflict between Syria and Iraq almost escalated into a war. The dam reduced the amount of water in the Euphrates flowing into Iraq so greatly that many Iraqi farmers were literally left high and dry.

They are our riv and we'll do wh

Kamran Inan, member
of the Turkish parliament,
in an interview with Neue
Zürcher Zeitung (NZZ)
in October 2003.

Source: The Helsinki
Rules on the Uses of the
Waters of International
Rivers adopted by the
International Law Associ-
ation at the fifty-second
conference, held at
Helsinki in August 1966.
Report of the Committee
on the Uses of the
Waters of International
Rivers (London, Interna-
tional Law Association,
1967) Article IV

Each basin stat its territory, to a and equitable sl uses of the wate national drainag

s, it is our water,
 we want with it.

is entitled, within
 easonable
 are in the beneficial
 s of an inter-
 basin.

Syria was willing to increase the quota of water only after massive military threats had been issued. But in the following unusually dry year, tension mounted once again when, to fill its own dam reservoirs, Syria reduced by three-fourths the volume of water it had previously agreed to release to Iraq.

In 1990, the never-ending conflict escalated once more, this time because Turkey, for an entire month, held back more than half the water in the Euphrates to fill its new Atatürk Dam reservoir, the centerpiece of the Southeastern Anatolia Development Project. This although Turkey had signed a provisional bilateral agreement three years earlier with Syria, guaranteeing it a much higher quota of at least 16 billion cubic meters of water per year. The Turkish government used occasional curbing of the waters of the Euphrates, as well as open threats, to show Syria how the agreement was to be understood—as a tool to coerce Syria into politically supporting Turkey on the Kurdish issue.

At the inauguration of the Atatürk Dam, Prime Minister Süleyman Demirel very clearly expressed Turkey's fundamental negotiating position: "This is a matter of sovereignty. This is our land. We have the right to do anything we like. The water resources are Turkey's. The oil resources are theirs [the Arabs']. We do not say we share their oil resources. They cannot say they share our water resources." He added, "Whoever is at the source has a right to it that no one can dispute."

Turkey's project and Syria's dam and irrigation projects make Iraq beholden to both countries. Although Iraq has its own larger water reserves, they meet only 40 percent of the country's total water needs. Iraq must rely on secure and stable water supplies from the Euphrates, and especially the Tigris, to irrigate two million hectares of farmland.

But the 1990 treaty does not guarantee a clearly fixed minimum amount of water. According to the treaty, Syria is willing to allow an average of 52 percent of the Euphrates' water to flow into Iraq, but the total amount of water depends on the amount of water Turkey allows into Syria in the first place. If in dry years or for political reasons Turkey keeps the flow of the Euphrates below the negotiated minimum, Syria covers its own needs first before allowing the remainder to flow into Iraq. The victims are, as always in such disputes, those farthest downstream—subsistence farmers and agricultural cooperatives in Iraq.

At the same time, occasional restriction of the Euphrates' flow also reduces the water supply available from the Tigris. The Tigris is by nature a much saltier river and must be mixed with water from the Euphrates to reduce salinity and make its water usable for irrigation. Because water from the Euphrates has been so scarce, more than 50 percent of total cultivated lands in Iraq today are highly salinized.

Should Turkey indeed build the highly controversial Ilisu Dam on the Tigris as part of the Southeastern Anatolia Project, hostilities could escalate again. Usable water resources in Iraq would be reduced by another 10 percent, and the Tigris river water flowing into the country would be even more polluted and salty because of its previous agricultural use in Turkey. Iraq would become more politically dependent than ever on Turkey and Syria.

Water conflicts are rarely just about water. Disputes over the distribution and quality of water resources are frequently combined with political motives. Few of them are resolved through actual military force. Of a total of 1,831 international conflicts over water listed in a study of the period between 1948 and 1998, only 37 involved military action, but 93 did involve the threat of massive military, economic, and diplomatic sanctions.

These figures have led several political scientists to conclude that water wars are little more than mythical. More progressive conflict analysts counter that, in practice, it makes no difference what means are used to decide a war. They argue that an expanded definition of war should include military as well as economic, political, and diplomatic instruments of power–this would more accurately reflect reality.

p. 478 Sanir Akhra, Dhaka, Bangladesh, 2006: The army distributes drinking water in response to protests against water shortages. Rafiqur Rahman/Reuters

Water and the Middle East conflict

Nowhere else in the world is the complexity of water conflicts as dramatic as in the Middle East, where political, ideological, strategic, financial, and technical problems are inseparably intertwined. The water issue here embodies everything at once—it is cause and result, purpose and instrument, target and weapon, claim and a means of pressure. The world perceives this ongoing fifty-year conflict mostly as a political one about Israel's right to exist, a Palestinian state, and the right of two million displaced Palestinian refugees to return to their home. It is also about strategic and military hegemony in one of the most explosive intersections between East and West. But the Arab-Israeli conflict is, last but not least, about water too.

The problem, so difficult to solve, is easy to describe. Israel, Jordan, and the Palestinian Authority are fighting over water resources that are not large enough to meet all the needs of the region's rapidly growing populations. It is a bitter struggle, as whoever controls the waters of the Jordan and the three large aquifers in the West Bank practically controls the region's entire natural and renewable water resources. Water scarcity poses an existential and equal threat to Israel and to the people in the future state of Palestine. The sources of half of the water Israel needs lie outside its own borders. The Palestinian population in the occupied territories has not had autonomous control over its own water resources for nearly forty years. Syria and Jordan are also affected because they lost part of their water resources to Israel in 1967(→ p. 506).

Israel's most important source of water is the Jordan River. Some 320 kilometers long, the Jordan is fed in its upper reaches by the Hasbani River flowing from Lebanon, the Banias River from Syria, and the smaller Dan, which is the only one whose headwaters are in Israel. In its lower reaches, the Jordan is fed by the Yarmouk River from the Golan Heights, the Zarka River, and a number of smaller rivers whose sources all lie in Jordan.

There were armed border conflicts between the Arab League and Israel as early as 1951. In part, they were triggered by Israel's plan to build a large national water pipeline to channel almost half of the Jordan's waters across Israeli territory to the Negev Desert. Jordan, one of the ten most arid countries in the world, responded in 1958 by building dams and the East Ghor Canal (known today as the King Abdullah Canal) to divert water from the Yarmouk, a Jordan River tributary that forms part of the borders between Syria, Jordan, and Israel, transporting the water for 90 kilometers across solely Jordanian territory, parallel to the Jordan River. The intention was to prevent Israel from using the Jordan's waters south of Lake Tiberias for its irrigation projects.

Syria's and Lebanon's plans to divert the Banias and the Hasbani, two of the Jordan's tributaries, directly into the Yarmouk, led nine years later to an escalation of the conflict. In the 1967 Six Day War, Israel occupied, among other things, Syria's Golan Heights. In one fell swoop, Israel thus gained control over all of the Jordan's tributaries above Lake Tiberias.

Ever since Israel in 1964 began diverting 400 to 500 million cubic meters of water each year through its National Water Carrier system, the Jordan river south of Lake Tiberias has been reduced to a mere stream completely polluted by wastewater. Its waters are not even fit for agricultural use. The victims of this rerouting are the inhabitants of the West Bank.

But ecological consequences have been devastating, particularly for the Dead Sea. Its water level has sunk by more than 17 meters in the last thirty years, reducing its original surface area by one-third. Water experts reckon that this inland sea, now 1,000 square kilometers in size, will turn into a salt desert in a few decades if no efficient countermeasures are taken, like building a pipeline to import water from the Red Sea or the Mediterranean.

During the 1967 Six Day War, Israel occupied the West Bank, thus securing, in addition to the Jordan, all Palestinian renewable groundwater sources, which provide more than 300 million cubic meters of water a year. In violation of the Geneva Convention and in utter disregard of all potential peace treaties, these sources were immediately declared to be non-transferable Israeli state property. A network of security zones and restrictive water allocation enforced by the Israeli military administration and the state-owned water company Mekorot mean that the two million Palestinians living in the West Bank have been able to use only one-fifth of their own groundwater since 1998. About three-fourths of the West Bank's groundwater is channeled directly to Israel. Mekorot supplies 140,000 Israeli settlers in the West Bank with 5 percent of the available water. Each inhabitant in these illegal settlements receives four times as much water as a native Palestinian.

In the meantime, this imbalance has shifted even more. Groundwater levels in many places have sunk greatly because of the excessive depletion of aquifers, causing many older and smaller wells to run dry. More than 200 Palestinian villages in the surroundings of Hebron, Nablus, and Jenin have no access to piped water, making them dependent on rainwater collected in cisterns and water from tanker trucks.

From this point of view, the safety barrier, which is there officially to protect Israeli settlements from Palestinian attack, is proving to be a water barrier. The route of the barrier, which sometimes extends deeply into Palestinian territory, runs along boundaries which Israeli hydrologists had already earmarked in "maps of water interests" as zones of strategic importance to Israel in the mid-1990s, before any intifada.

The situation in the Gaza Strip is even more precarious, where water supply is practically completely dependent on groundwater reserves.

Although the Palestinian population is twenty-six times larger than it was 1948, the amount of water available to it has remained the same.

Conditions worsen there from year to year. Most of the water supply system in the Gaza Strip has been destroyed and remaining pipes leak in many places. The excessive extraction of water from aquifers has caused groundwater tables to sink drastically here too. In areas where groundwater sources are near the sea, seawater has seeped in, salinizing groundwater to such an extent that it can no longer be used for drinking. Other groundwater reserves in the Gaza Strip are replenished by drainage from neighboring Israeli farmland. The water is so contaminated with pesticides that poor water quality has led to widespread health problems for the population living in the Gaza Strip. UN experts fear that the drinking water supply in the Gaza Strip will completely collapse in the next ten to fifteen years.

The American government worked out a plan in the mid-1950s to fairly divide water reserves in the whole Jordan Basin. Named after mediator Eric Johnston, the plan was tacitly accepted by Israel and the Arab League as a basis for negotiations until the 1967 Six Day War. For political reasons, neither party ever ratified the Johnston Plan.

All other attempts to solve the water problem in the region failed. The parties involved refused to engage in multilateral negotiations that would have included all affected groups, and that could have resulted in a binding treaty on this complex issue. While there have been no negotiations whatsoever with Syria and Lebanon, Israel and Jordan did reach bilateral agreement in a peace treaty signed on October 26, 1994, which largely recognized the already existing allocation of Jordan River waters.

In contrast, all Israeli governments to date have refused to negotiate a binding agreement with the Palestinians. Although water allocation has always been one of the most controversial core issues in negotiations, Israel made only non-binding concessions in the Oslo Accords of September 13, 1993, the Gaza-Jericho Agreement of May 4, 1994, and Oslo II of September 28, 1995. Israel generally acknowledged Palestinian water rights in the West Bank in the Oslo II Treaty, but exact allocations and their technical implementation will not be determined until negotiations regarding Palestine's definitive status take place.

Despite international principles of law, Israel today uses its power as an occupying force to push through preliminary decisions to its own advantage. A case in point is Israel's declaration of a so-called security zone along the Jordan River, which it doesn't want to include in any peace talks. This effectively denies Palestinians the right to participate in decisions concerning the waters of the Jordan River. Israeli negotiating delegations have also managed to get Palestinian groundwater resources defined in agreements as resources whose use will be determined by a water committee set up after a peace treaty is concluded. In this way, Israel has already secured itself the right to veto any decisions regarding the use of water resources, even in future Palestine. This is a right Israel however denies Palestinians in the Gaza Strip, where the Israeli water company Mekorot will continue to be the sole water supplier.

But a certain procedure established in the Oslo II Treaty has even graver consequences. "Existing" as well as "new, additional" water resources are to be negotiated separately. Water from desalination plants, for instance, is defined as a new, additional water resource. Water experts believe this differentiation doesn't make any sense because both types of water resources are inseparably linked and can be effectively negotiated only as a whole. If a larger share of "existing" water from the Jordan River and the West Bank are channeled to Israel's coastal regions, then an equal amount of desalinated water would have to be pumped from Israel's coastal regions back to the West Bank to balance the ensuing water deficit in Palestine. This is technically feasible, although it would be an immensely costly project involving "two-lane" water transport. More importantly, it would also make Palestine entirely dependent on Israel politically, not only because desalination plants would be in Israeli territory, but also because water pipelines would cross Israeli territory, giving Israel complete control over Palestinian water resources.

This is a vicious circle, hard to break. Israel is not prepared to resolve conflicts over water with Palestine until a peace treaty has been signed, and Palestine can agree to a peace treaty only after these problems have been solved.

p. 486 Protest by Mazahua women against the dams planned on their land. They are designed to supply Mexico City with water. Mexico City, 2005. Daniel Aguilar/Reuters

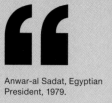

Anwar-al Sadat, Egyptian President, 1979.

1979: The only t would ever fight is water.

Boutros Boutros-Ghali, Egyptian foreign minister (later UN General Secretary), 1987

1987: The next w will be fought ov of the Nile.

Egypt's minister for water, speaking in 2003, when Kenya put in question Egypt's allocation of Nile water.

2003: This is an

ng that Egypt
a war over again

ar in our region
r the waters

ct of war.

The need for international water legislation

Water does not comply with international law. The principle according to which a state is the sole owner of all its natural resources is hard to apply to international waters or groundwater reserves (aquifers). The natural water cycle doesn't care about political borders, and this forces neighboring states to enter negotiations on how they want to share and use their common waters.

There are still hardly any binding international rules for this process. Nor is there any universal consensus in the form of international legislation on water which would establish the principles and norms for using and allocating international waters or delineate the rights and obligations of neighboring states. This is a paradoxical situation. Since fundamental laws are either missing or so vaguely worded that they leave a lot of room for interpretation, the tools which could be used to resolve conflicts over water themselves become subject to dispute. It is indeed conceivable that negotiating or conflicting parties would therefore attempt to influence the selection of rules for the negotiating game to such an extent that they could determine the solution of the conflict in advance.

Admittedly, the negotiating positions of parties as a rule start off being asymmetric. Nature has already generally given upstream states the upper hand, but not always—economic and military power, security interests, and alleged historic rights are also crucial. But an unequal balance of power is often more decisive for the outcome of negotiations than rational arguments or possible pragmatic solutions.

In the 1895 dispute between the United States and Mexico over the use of the Rio Grande, Judson Harmon, the attorney general of the United States at the time, expressed his belief that every country held the exclusive right of disposal over all waters in its territory. But the Harmon Doctrine, as it came to be known, obviously so offended any moral sense of justice that the United States itself, eleven years later, granted its neighbor what it considered to be an appropriate share of the Rio Grande's waters.

The Unites States was far less conciliatory when it came to the more expensive question of what levels of water pollution Mexico would have to accept. As long ago as 1962, Mexico started protesting against the completely unacceptable quality of water in the Colorado River. Intensive irrigation in the United States had caused the river's naturally high salinity to triple by the early 1960s. Crop yields in the irrigated fields of

Soghran Bibi, a middle-aged resident of the Lyari neighborhood of Karachi in Pakistan, one of the areas where riots broke out in protest against acute water shortages.
Gulf News, 30 June 2003

In the localities give water to the their cars. But ir not received wa days. Shouldn't throw stones?

f the rich, people

r lawns and wash

my house I have

er for the last four

burn tyres and

the Mexicali valley were decreasing from year to year. Polluted with numerous chemicals, river water could no longer be used for drinking water anyway.

The United States took its time. Not until 1974, one-and-a-half decades later, did it declare itself willing, in a treaty, to take on a certain amount of responsibility for the quality of Colorado River water. But it still took another eighteen years for the U.S. to start operating a desalination plant in Yuma, Arizona, just above the Mexican border. The plant would have greatly improved water quality if it hadn't shut down a year later because of constant breakdowns and unexpectedly high costs. There have been lukewarm discussions since then about whether the Yuma plant should be renovated or whether other, less costly solutions should be sought.

Negotiations have been further complicated by the fact that the states of Nevada, Arizona, and California are themselves fiercely arguing about the use of the Colorado River and how to keep it clean. A solution to the conflict between the United States and Mexico is still not in sight, even after decades of negotiations. Mexico is hardly in a position to put pressure on how the U.S. is managing this problem.

Following the usual script, downstream states are the ones who usually threaten economic sanctions or the use of military force. In 1979, Egypt's president Anwar Sadat warned, "If Ethiopia takes any action to block the Nile waters, there will be no alternative for us but to use force." Egypt is on the lower reaches of the Nile and 97 percent of its water supply depends on the river. In 1929 and 1959, Egypt had negotiated very advantageous treaties with Sudan, thanks to more or less high military pressure and a few air raids. Sudan guaranteed Egypt more than two thirds of the Nile's total water volume (and an even higher share during the summer months) and granted Egypt the right to veto any decisions on the future allocation of Nile waters.

In their bilateral negotiations, Egypt and Sudan conveniently forgot that four-fifths of the Nile's waters do not originate in Sudan, but in the mountains of Ethiopia and that most of the rest originates in Uganda and Burundi. Apparently Egypt and Sudan thought these three real source countries of the Nile were too unimportant for their needs to be taken into consideration too.

When Ethiopia vaguely expressed its needs late in the 1970s, and in the early 1990s clearly voiced its intention of using water from the Blue Nile for its own agriculture, Egypt was not the least bit willing to enter negotiations and threatened Ethiopia with war. This in spite of studies by American hydrologists that showed Ethiopia's plans could indeed have a positive effect on all the states sharing the Nile if they were willing to cooperate. Ethiopia would increase its urgently needed irrigated areas almost twenty-fold from less than 190,000 hectares to 3.7 million hectares. Sudan would increase its crop yield even more since the water level of the Blue Nile would be regulated thanks to Ethiopian dams, providing Sudan with enough water even in dry summer months for irrigation. Last but not least, Egypt would not even have to make do with less water. The water level of the Nile, kept even throughout the year, would allow engineers to massively reduce the amount of water in the reservoir behind Egypt's Aswan Dam. Calculations revealed that this would lead to much less water being lost from the reservoir through evaporation, the amount of water saved almost exactly equaling the share of water retained by Ethiopia for its own irrigation projects. A win-win situation for all three countries.

In spite of its unbending position, Egypt did repeatedly initiate negotiations and cooperation conferences that led to hardly any results until a few years ago. It wasn't until the Nile Basin Initiative was launched in 1999 that the countries involved were treated as equal partners for the first time and agreed that a fair division of the Nile's waters would enable sustainable development for all participating states. A strategic program of action is supposed to create the conditions for integrated water management in the whole catchment area of the Nile.

In the Mekong River Basin in Southeast Asia, Laos, Thailand, Cambodia, and Vietnam have been trying for almost fifty years, despite violent political turmoil, to peacefully coordinate the use of the Mekong's waters (→ p. 507). A first organization, the Mekong Committee, was set up as long ago as 1957 at the initiative of the United States, not least to turn these four countries into a bulwark against communist China. The present Mekong River Commission (MRC), founded in 1995, must still face its acid test regarding the integration of China's and Myanmar's interests.

Paradoxically, the war in Vietnam and later civil war and the political isolation of the governments of Cambodia and Burma protected the Mekong from being overly used or industrially exploited. These wars had the effect of keeping Vietnam, Cambodia, and Burma in an extreme state of poverty. About 80 percent of the people in the Mekong Basin have subsisted for centuries on seemingly medieval methods of agriculture and fishery. During the rainy season, the Mekong River floods, fertilizes, and soaks large areas of farmland, from the steep slopes of the mountainous regions of Laos and Cambodia, to the flat estuary in Vietnam. Between the rainy seasons, rice, grains, and vegetables are cultivated, providing food for about 300 million people. The area also yields 2 to 3 million tons of fish a year. Tonle Sap, a branch of the Mekong, swells to five times its normal size during the monsoon season, becoming the largest lake in Southeast Asia. The livelihoods of 1.2 million families, 2 to 3 million people living on Tonle Sap alone, depend almost exclusively on fishery.

But this could change drastically in coming years, mostly because China wants to intensively use the waters of the upper Mekong to generate electricity (→ pp. 365). Two large hydroelectric dams are already operating on its upper reaches, and two more are currently under construction. There are plans for a total of eight dams.

Furthermore, China wants to make the river navigable for larger riverboats all year long over a 880-kilometer stretch between the cities of Simao in Yunnan province and Luang Brabang in Thailand. If these projects go through, they will permanently upset the river's water balance farther downstream. Flooding that is crucial for cultivating rice and naturally fertilizing fields will no longer occur. As a regulated river, the Mekong would not be thoroughly flushed every year, important wetlands would be destroyed, and huge fishing grounds would disappear.

Since the 1990s, Thailand, Laos, and Vietnam have begun building larger dams and irrigation systems. Two Laotian hydroelectric dams on the Mekong tributaries of Nam Theun and Se Kong are already in operation, and four more are currently being built. In 1998, Vietnam completed a large dam on the Sesan River, the Mekong's largest tributary, and there are plans for three more dams. Inhabitants are already feeling the consequences of this regulation of the Mekong. Fish stocks downstream from dams are declining, and the crop yield of 50,000 Cambodian farmers diminishes every year.

The four members of the Mekong River Commission signed a technical agreement with China and Myanmar for the first time in April 2002. It dealt with less contentious issues such as improving the exchange of information on flood forecasts. In August 2004, the MRC agreed with China and Myanmar to improve general communication by consulting regularly with each other. This includes discussing each country's plans for using the Mekong's waters and China's plans for making the river's upper course navigable. Finally, a flood management plan and an environmental program are to be initiated.

The commission hopes this could be a first step towards a future treaty on joint integrated river management. But skeptical diplomats anticipate the possibility that China's greatly superior economic power, which all other countries in the region are more or less dependent on, could lead to new coalitions and strategic alliances within the commission which would break down the old MRC states' present willingness to cooperate. Until now, their joint resistance to China's dam projects drowned out, so to speak, these four countries' internal conflicts. China could now attempt to pit member countries against each other to its own advantage.

It is no coincidence that the most promising approaches towards international regulation have been developed in Europe. Europe never had to deal with such existential issues as the allocation of limited water resources, only with preventing water pollution. Wealthy European countries already have an excellent water infrastructure and the cost of sanitation measures simply doesn't play the important role it does in poorer developing countries. Added to that, democratically structured countries within Europe enter into negotiations with countries sharing similar cultural, political, and social values—an imbalance in economic and military power in these countries is negligible.

While countries along the Rhine were able to agree on protective measures against pollution in the mid-1950s, those along the Danube weren't able to do so until the mid-1990s. The struggle between different social systems in states along the Danube prevented cooperation between the capitalist countries of Germany and Austria and the majority of former communist countries—until the collapse of former eastern bloc regimes.

In contrast, Switzerland, France, Germany, Luxembourg, and Nether-lands, who share the Rhine, set up the International Commission for the Protection of the Rhine (ICPR) in 1950. Europe's third longest river, the Rhine flows through one of the world's core industrial regions. Since the beginning of industrialization, it has been one of the continent's most heavily contaminated bodies of water, earning an inglorious reputation as Europe's sewer from early on.

In 1963, these five states signed the Treaty of Bern, which the European Economic Community, now the European Union, also signed in 1976. It provides an internationally binding framework for specific measures. It was followed by several lesser treaties, and in 1987, after the Sandoz disaster (→ pp. 302), the Rhine Action Program for Ecological Rehabilitation (RAP) for the first time set some general goals. By 2000, animal spe-cies that had disappeared from the Rhine were to be reintroduced, the Rhine's waters were to remain suitable for drinking, and hazardous sub-stances in the river's sediment were to be significantly reduced.

The ICPR took stock in 2000 and found that many listed harmful chemi-cals had been reduced by 70 to 100 percent, 95 percent of industrial operations and communities had been connected to water treatment plants, and nearly all of the Rhine's domestic fish species were back.

The new Convention for the Protection of the Rhine entered into force in 2003 and prescribed for the first time the comprehensive protection of the Rhine's entire ecosystem. With this convention, the ICPR is aiming for much more than technical river management. According to the convention, the Rhine ecosystem encompasses everything that interacts with the river—surrounding land, floodplains which are to be recovered, and groundwater. For the first time, nature is placed on an equal footing with humans, and with commerce and industry. This convention could certainly have a determining influence in the design of international water law.

p. 496 UNICEF supplies drinking water to Iraq, 2003. Atta Kenare/Keystone

The commercialization of a human right

Is access to water a human right or just a human need? As strange
and obscure as this question may sound to the layperson, it has
been the subject of debate for more than two decades at meetings
of experts, international scientific congresses, United Nations conferences, and even at world summits, without any chance of agreement
being found.

The argument is more than just a technicality of language. At the heart
of the debate is the question of which strategy to use to most effectively
combat the global water crisis. If access to water is a human right, all
states and the world community at large are obliged to guarantee access
to safe drinking water to all, regardless of where they come from, how
they live and how much money they have. In contrast, if access to water
is merely a human need, the same conditions apply as to any other
commercial commodity. Even though the answer to this question has
enormous consequences, the issue was hardly contentious until the 1970s.
Even at the 1977 UN Conference on Water at Mar del Plata, for instance,
when governments promised to guarantee everyone safe access
to clean drinking water and adequate sanitary facilities by 2000, hardly
anyone questioned the right of access to water. There was a silent
assumption that this was consensus.

The promise made at Mar del Plata gave rise to many development projects
and community initiatives, yet the goal, while optimistic, was far from
attainable and bound to fail. It was not based on a country-by-country
analysis of specific situations regarding water but was rather an expression of wishful thinking and diplomatic rhetoric. Instead of looking for
options that were inexpensive and geared to the real capabilities of
developing countries, donor countries and the two most important money-
lending organizations, the World Bank and the International Monetary
Fund (IMF), banked on expensive, centralized solutions and complex
western technology. This approach later contributed to many recipient
countries being hopelessly unable to repay their loans, plunging them
heavily into debt.

When failure became evident, the prime concern of the World Bank and
the IMF was then to tighten the conditions under which they granted
loans, not to change their unrealistic technological objectives. To relieve
heavily indebted state budgets in developing countries, these lending
institutions decided to create investment incentives for large international construction and supply businesses, attaching a number of conditions
to the granting of new loans. State and municipal water utilities were to
be privatized and opened to investment from foreign businesses. Subsidies

that distorted competition were to be reduced and water fees were to be structured to cover effective costs and reap attractive profits.

This strategy of commercialization, designed by international financial institutions, indeed broke the silent consensus that access to water was a human right. Without the question being settled, these institutions turned water into a commercial commodity so that access to water was in practice limited to those consumers who could afford to pay water fees.

This new practice was reflected, although rather modestly at first, in discussion papers circulating at international conferences in the 1980s. In the Dublin Statement on Water and Sustainable Development, issued at the 1992 International Conference on Water and the Environment in Dublin, which later influenced the wording of Agenda 21, water experts from one hundred national governments and national and international non-governmental organizations (NGOs) agreed that "water should be recognized as an economic good." The intention was to make sure that water as a valuable and scarce commodity would be used more carefully and economically, especially in industry, business, and agriculture. But the Dublin Statement at the same time expressed the need that this should happen without limiting the right of access to safe drinking water. In practice, these two objectives proved to be more and more contradictory, and continuing debate on how to reconcile them was hardly fruitful.

In the course of the 1990s, it became clear that the imposition of privatization as desired by the World Bank and several bilateral development institutions did not lead to the success expected. Water supply hardly improved in affected developing countries. International water companies failed to make anticipated profits and therefore often limited the work contracted to selected services. In many cities, large international water suppliers withdrew from their contracts and reduced their commitment in developing countries by half or more (→ pp. 458).

Voices of criticism grew louder, even at the World Bank, without this leading to a basic reconsideration of the strategy being pursued. On the contrary, the World Bank reacted by making loan conditions even stricter. To keep the private sector involved, incentives for investment were hiked up again. International money-lending institutions wanted private investors to be given more scope for action, and for their responsibilities towards the public to be reduced, their financial risks limited, and their profit expectations better safeguarded.

In the meantime, other international organizations were increasingly looking at the issue of water. Experts around the world had agreed for a long time that water was part of a more comprehensive and far more complex set of problems. They knew that hunger, poverty, health, water, environment, and climate had such an intense effect on one another that any one of these problems could not be dealt with separately from any other.

Over the years, the issue of water was discussed at a great number of international conferences and forums, in vastly diverse ways and often with opposing results. At conferences focusing on environment, food, or health issues as global problems facing humanity, experts clearly distanced themselves from purely economically driven plans for deregulation and privatization. In their basic declarations, resolutions, and recommendations, United Nations organizations like the Food and Agriculture Organization (FAO), the World Health Organization (WHO), UNESCO, the UN Environment Program (UNEP) and the UN Development Program (UNDP) to a great extent stood by their perception that access to water, similar to people's entitlement to a healthy environment, was a human right and therefore the responsibility of governments and the world community.

In contrast, wherever conferences centered on water management as an industry, or focused on the establishment and operation of water utilities, a nearly unanimous position was taken up in which water is primarily an economic commodity and its most efficient use should be left to the "invisible hand of the market." This position was also supported by the World Water Council and the Global Water Partnership, an action-oriented network, two supranational organizations founded in 1996 at the initiative of the World Bank and considerably influenced by it. The principles of future global water policy defined by these two institutions were much more extensive and influential than those expressed at United Nations conferences. The privatization position was fortified by negotiations conducted within the World Trade Organization to set up the General Agreement on Trade in Services (GATS), in which water (with a few exceptions) is supposed to be classified as a commercial commodity.

At the 2001 International Conference on Freshwater in Bonn, which served to prepare for the 2002 UN World Summit on Sustainable Development in Johannesburg, these opposing strategies collided head-on with each other once again. In particular, numerous NGOs and unions heavily criticized the strategy of privatization, as did several government representatives, mostly from developing countries. They used numerous examples to document the failure of privatization, proving that water companies were obviously uninterested in providing the poorest population groups with water wherever it was unprofitable. They also criticized the inadequate monitoring and regulation exercised by governments and local authorities.

Although these critical groups, calling for a declaration that access to water must be an inalienable human right, were not successful, they were able to get a compromise wording into the final document: "The primary responsibility for ensuring the sustainable and equitable management of water resources rests with the governments." The document also calls for an independent examination of present experience with private-sector participation, and asks bilateral donor agencies and the World Bank to stop making privatization a prerequisite for granting loans.

Sadly, these demands were barely heard at the Johannesburg Summit. Apart from issuing the Johannesburg Declaration, a political declaration of principle, the summit focused on adopting a program of action which was supposed to implement the millennium goals. But instead of reviewing what progress had been made since the 1992 Earth Summit in Rio de

Janeiro and resolving to adopt further specific measures within clearly defined time frames, as had been originally intended, participants at the Johannesburg conference limited their activities to declaring they would join numerous non-committal partnership initiatives. The European Union, for example, announced its Water for Life initiative, in which it is supposed to work closely together with European water companies to improve water supply in Eastern Europe, central Asia and Africa. The United States Agency for International Development announced a similar initiative for western Africa, and the Managing Water for African Cities program is supposed to be carried out by UN-HABITAT and UNEP in cooperation with national governments, the private sector, and civil society organizations. But the private sector and northern donor countries did not make any binding promises to finance these projects.

Although the Johannesburg Summit did not deal explicitly with clarifying the question of whether water is a human right or merely a trade commodity, the conference unmistakably took a stand in favor of commercialization. The concept of joint public-private partnerships had never before been as emphatically propagated as at the summit meeting in Johannesburg. There are more than a hundred references in the final documents to the fact that projects should in no case be carried out "without prejudging the outcome of the World Trade Organization negotiations." Delegates thus left the clarification of the human rights question to an institution whose main task is to liberalize and deregulate world trade, and which strives in GATS negotiations for far-reaching privatization, even of public services.

(Likewise, supporters of private, commercially driven solutions, led by the United States and the European Union, had their own way on the question of financing development projects. The final documents state that development financing must be oriented along the guidelines of the Monterrey Consensus. This agreement was adopted at the UN's 2003 International Conference on Financing for Development and is a slightly modified version of the 1989 Washington Consensus, the most important paper on the principles of neoliberal globalization. Thus, the Monterrey Consensus also emphasizes the central role of private capital in financing development projects and calls on developing countries to keep improving their framework conditions for supporting and protecting direct foreign investment.)

At Johannesburg, critics probably rightly accused northern donor countries of asserting the interests of their own large construction and service industries. Critics believed that enhancing the private sector would seriously limit the scope of the public hand. This is because WTO trade regulations can be enforced through legal action, possibly allowing private water suppliers to exert decisive influence on the framework conditions of projects, even against the will of governments. This holds true especially wherever the legal system relevant to such matters is inadequate in developing and emerging countries, and the authorities needed for administration and inspection are not present.

Those concerns that go beyond single projects and require comprehensive regulating, such as protecting the environment, conserving wetlands, maintaining biodiversity, and avoiding negative social effects, would fall by the wayside. Because these objectives usually cost money, they would be left out of the calculation, handed over to the public, or simply ignored.

Of course, municipal authorities, representatives of the private sector, aid organizations, and local NGOs in many places have long since successfully worked together on many projects in spite of all their differences. Although general goals and basic issues are continually argued over at major international conferences, in recent years more and more conference workshops have looked at pragmatic approaches and the specific needs of local authorities. The International Council for Local Environmental Initiatives, a union of nearly 500 towns and municipalities in forty-three countries, was already engaging in numerous activities at the Johannesburg Summit to draw attention to the specific needs of local administrations. Together with the United Nations organizations UNEP, UN-HABITAT, and WHO, it founded an own initiative called Local Action 21. The motto of the World Water Council's most recent conference, the 4th World Water Forum, held in Mexico City in March 2006, was "Local Action for a Global Challenge."

While the right to water still remains a hotly debated issue, it is not just behind the scenes that hardened attitudes are beginning to soften. This is certainly a consequence of civil society groups, environmental associations, development organizations, aid societies and local initiatives having had at least limited access to many conferences.

But probably the more decisive factor is that current strategies and efforts clearly will not suffice to meet the millennium goals. This was also indicated in the 2nd UN World Water Development Report presented in 2006 at the World Water Forum in Mexico. The report says that the privatization of water supply and sanitation facilities, as well as the decentralization of decision-making structures, have not met the hopes placed in these approaches. The report identifies major shortcomings in handling water, particularly at the important level of local administration. It also testifies that private investment in poor developing countries has clearly lagged behind expectations, and that development loans from international institutions like the World Bank have also clearly stagnated.

The ministerial conference at the World Water Forum rejected by a large majority the demand to make access to water a human right. Ministers limited their statement merely to pointing out the "critical importance of water for all aspects of sustainable development, including poverty and hunger eradication." The compromise wording in this case was: "Water is the guarantee of life for all the world's people." But delegates reporting on negotiations said that nearly all governments had replied in the affirmative to the question of the right to water. Ministers explained that the main reason for rejecting this concept was that as a right it was difficult to translate into action and could lead to legal complications on both national and international levels.

A few days after the 2006 World Water Forum ended, the World Bank, which until now has been one of the most resolute advocates of privatization, indicated it may change its strategy. In an interview with the German daily newspaper Frankfurter Rundschau, Katherine Sierra, a vice president at the World Bank, said: "The World Bank's commitment to the private sector in the 1990s had an ideological nature." She added that the World Bank would now work together with representatives from across the whole political spectrum. Local municipalities would have to make pragmatic decisions themselves on whether they preferred public or private operators and suppliers.

Important international water conferences 1972–2006

1972	UN Conference on the Human Environment	Stockholm	Founding of the United Nations Environment Programme (UNEP)
1977	UN Conference on Water	Mar del Plata	Assessment of water resources, water use and efficiency. Mar del Plata Action Plan
1981–1991	International Drinking Water and Sanitation Decade		The 1977 United Nations Conference on Water set up an International Drinking Water Decade, 1981–1990. Its aim was to make access to clean drinking water available across the world.
1990	UNDP Global Consultation on Safe Water and Sanitation for the 90s	New Delhi	"Some for all rather than more for some." The New Delhi Statement is an appeal to all nations for concerted action to enable people to obtain two of the most basic human needs—safe drinking water and environmental sanitation.
1990	UNICEF World Summit for Children	New York	Declaration on the Survival, Protection and Development of Children
1992	UN International Conference on Water and the Environment	Dublin	The Dublin Statement on Water and Sustainable Development.
1992	UN Conference on Environment and Development (UNCED Earth Summit)	Rio de Janeiro	Rio Declaration on Environment and Development, a non-binding statement (voluntary implementation from 1994 to 2000). Goal: to reduce CO_2 levels emissions by the year 2000 to the levels of 1990 /Agenda 21
1994	Ministerial Conference on Drinking Water Supply and Sanitation	Noordwijk	Action program to assign high priority to Chapter 18 of Agenda 21.
1994	UN International Conference on Population and Development	Cairo	Programme of Action
1995	UN World Summit for Social Development	Copenhagen	Copenhagen Declaration on Social Development
1995	UN Fourth World Conference on Women	Beijing	Beijing Declaration and Platform for Action
1996	UN Conference on Human Settlements (Habitat II)	Istanbul	The Habitat Agenda
1996	UN World Food Summit	Rome	Rome Declaration on Food Security
1997	1st World Water Forum	Marrakech (World Water Council)	Marrakech Declaration recommends action to recognize the basic human needs to have access to clean water and sanitation, to establish an effective mechanism for management of shared waters, to support and preserve ecosystems, to encourage the efficient use of water, to address gender equity issues in water use, and to encourage partnership between the members of civil society and governments.
1998	UNESCO Conference on Water and Sustainable Development	Paris	Paris Declaration
2000	2nd World Water Forum	The Hague (World Water Council)	World Water Vision: Making Water Everybody's Business.
2000	Ministerial Conference on Water Security in the 21st Century	The Hague	Seven challenges: meeting basic needs, securing the food supply, protecting ecosystems, sharing water resources, managing risks, valuing water, governing water wisely.
2000	55th Session of the General Assembly of the UN		UN Millennium Declaration

2001	Third International Conference on Groundwater Quality	Sheffield	
2001	Special Session of the UN General Assembly for an Overall Review and Appraisal of the Implementation of the Habitat Agenda, Istanbul+5	New York	
2001	2nd IWA World Water Congress	Berlin (International Water Association, IWA)	
2001	UN Conference on Freshwater	Bonn	Ministerial Declaration / Recommendations for action
2002	3rd IWA World Water Congress	Melbourne	
2002	International World Bank Conference on Water and Sanitation Service in Small Towns and Multi-Village Schemes	Addis Abeba	
2002	Third World Social Forum	Porto Alegre	Water is a social good that should not be bought and sold.
2002	UN World Summit on Sustainable Development, Rio+10	Johannesburg	The number of people without access to safe drinking water and sanitation is to be halved by 2015.
2003	UNDESA/UNESCO International Year of Freshwater		
2003	WHO/IWA Water Safety Conference: Risk Management Strategies for Drinking Water	Berlin	
2003	3rd World Water Forum	Kyoto (World Water Council)	1st edition of the World Water Development Report/ Ministerial Declaration. By 2025 everybody is to know the importance of hygiene and have access to an adequate safe water supply and sanitation.
2004	Third Forum on Global Development Policy: Water – human right or commodity?	Berlin (Heinrich Böll Foundation)	
2004	UNEP/GPA Global H_2O Partnership Conference	Cairns	
2004	4th IWA World Water Congress	Marrakech	
2006	4th World Water Forum	Mexico (World Water Council)	See chapter Water and Power, The commercialization of a human right (p. 498)

Israel/Palestine

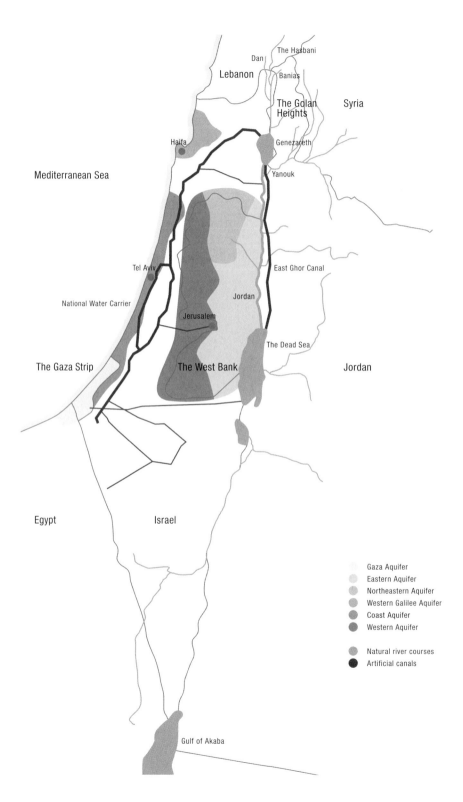

Lebanon

The Hasbani

Dan

Banias

The Golan
Heights

Syria

Haifa

Genezareth

Mediterranean Sea

Yanouk

Tel Aviv

East Ghor Canal

National Water Carrier

Jordan

Jerusalem

The Gaza Strip

The Dead Sea

The West Bank

Jordan

Egypt

Israel

Gulf of Akaba

Gaza Aquifer
Eastern Aquifer
Northeastern Aquifer
Western Galilee Aquifer
Coast Aquifer
Western Aquifer

Natural river courses
Artificial canals

Dams on the Mekong and the Salween Rivers

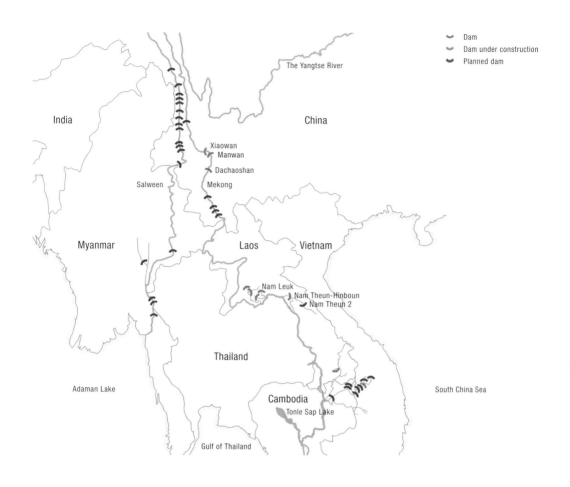

Dam
Dam under construction
Planned dam

The Yangtse River

India

China

Xiaowan
Manwan

Dachaoshan

Salween

Mekong

Myanmar

Laos

Vietnam

Nam Leuk
Nam Theun-Hinboun
Nam Theun 2

Thailand

Adaman Lake

Cambodia
Tonle Sap Lake

South China Sea

Gulf of Thailand

Southeast Anatolia project

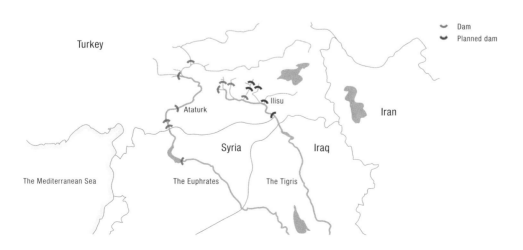

Dam
Planned dam

Turkey

Ataturk

Ilisu

Iran

Syria

Iraq

The Mediterranean Sea

The Euphrates

The Tigris

WATER BELONGS TO US ALL.

Water belongs to us all – an appeal

It's a paradox: politicians, lawyers, and economists at numerous international conferences have been arguing for decades over an issue that can't genuinely be a question of dispute any more, even for them. Everyone agrees that water doesn't belong to anyone and therefore belongs to us all. Only one thing contradicts this point of view–reality.

The discrepancy between theory and practice goes back a long way. Before legal systems were formally instituted, ancient peoples imbued water with holiness, substantiating its importance with the most authoritative of values. Whatever belongs to the gods can't belong to individual human beings. At the same time, however, rulers and kings selfishly defended their waters, keeping them from others, if necessary even from their own people.

The Enlightenment did away with the idea that water was sacred. Knowledge no longer centered on metaphysics or meaning, but on usefulness; it became pragmatic. "The human mind, which overcomes superstition, is to hold sway over a disenchanted nature," said the most rigorous of analysts Theodor Adorno and Max Horkheimer in *The Dialectic of Enlightenment*. "What men want to learn from nature is how to use it in order wholly to dominate it and other men." Water as H_2O, merely a substance, lost the mysterious powers once attributed to it. It became a commodity.

The modern era has demystified water. Chemists, physicists, and biologists have discovered how water functions, and hydrologists, marine researchers, and climatologists have brought to light the complex mechanics of water cycles. As expected, scientists have not found holiness within molecules, nuclei, and their electron shells, but they have discovered properties that accord water its unique place among all substances. Nothing moves without water. Water keeps everything going, from the largest global processes that determine Earth's composition and climate, to the smallest chemical and biological events, without which life or change would be impossible.

The demystification of nature has also done away with a second myth. The ancient threat was that the gods would punish whoever abused nature. But who can punish whom for what when there are no gods to do that, and water is nothing but a profane consumer item? Modern scientists have only gradually decoded the real message behind the mythical threat. If we fail to take good care of water, we will ultimately punish ourselves. If used at will like a consumer good, water is rendered foul and unsuitable for any purpose.

In our secularized world, only the concept of human rights bears the same authoritative power as the religious myths of prehistory. Human rights define the inalienable prerequisites for living in dignity, and all humans are entitled to them. Indeed, everything speaks for one of these rights being access to water. Without water there is no human life. Without water there would be no food or clothing, no nature, no culture.

As obvious as this cognizance may be, the reality is that access to water is in practice seldom exercised as a human right. One-sixth of humanity still has no reliable access to enough clean drinking water. Every third person is unable to look after his or her most basic needs with regard to hygiene. Half of the people in Asia and Africa, Earth's largest continents, live in degrading and inhumane circumstances and often don't have enough water. If scientists' forecasts are correct, these numbers will soon be much higher. Scientists estimate that in thirty to forty years, half of humanity will live in the poor slums of overpopulated megalopolises—precisely where water is already most scarce.

The polemic maxim that "human rights are no help here, we can't even wash our hands with them" sadly seems to be true. Access to water calls for building infrastructures, but the cost of these far exceed the financial means of any individuals or businesses, particularly in the industrialized and urbanized world. Gérard Mestrallet, head of the Suez-Ondéo international water group, once made a now familiar quip: "God provided the water but not the pipes." For all its cynicism, his remark indeed hit directly on the central question of water supply. Who will cover the enormous costs of water management in the industrial age? Who will make sure water gets to where it's needed? Who bears responsibility and covers the cost of treating polluted water so that it doesn't endanger the natural environment or humans? Who is most efficient at solving these problems at a reasonable price?

The broad consensus that reigned until a few decades ago was that water supply must be a common responsibility borne by municipalities and states since state institutions are obliged to look after the common good. Only these institutions are legitimized and have the authority to enforce their obligations towards the citizenry. Only they can guarantee fair distribution through taxation and levying fees, acting in solidarity to even out differences between poor and rich population groups and between favored and disadvantaged regions.

Liberal economic theorists in the 1980s posed a radical alternative to this concept. Believing state administration to be bureaucratic, sluggish, inefficient, and in many parts of the world corrupt, theorists maintained that the "invisible hand" of the market (Adam Smith) was best able to solve humanity's material problems. The only guarantee that scarce water could be distributed sensibly and at low cost, in accordance with the most pressing needs, would be by handling it as an economic good, its price determined by the incorruptible market of supply and demand.

This theory claimed that the most efficient way to optimize the free play of market forces was to encourage completely unrestricted competition between private water suppliers. Advocates of a globalized, deregulated, and liberalized world economy saw the state as the greatest obstacle to this open competition. They maintained that the state's specific interest in protecting domestic markets and in exerting economic power abroad distorted the rules of the game. It wasn't a coincidence that neoliberal theory linked its call for "more freedom" with a plea for "less state."

But this economic theory is based on a premise that seems to be rather unrealistic in face of today's actual situation. It presumes the existence of a *homo oeconomicus* who naturally has money at his or her disposal. It doesn't take into account the fact that millions of people don't have any money, yet are interested in staying alive.

To date there is no conclusive evidence that this recipe for deregulation, liberalization, and privatization really works in developing countries. Reputable economists, among them Amartya Sen, Paul Krugman, Joseph Stieglitz, even Paul Samuelson, the doyen of modern economics, indeed none of them opponents of a free market economy, doubt that globalization, privatization, and free trade can really be engines of prosperity for poor countries. At least not as long as "self-destructive forces" (Stieglitz) and extreme inequality in the distribution of power and wealth are not contained and corrected.

This is because globalization, in particular the deregulation of financial markets, has since led to a marked loss in political options. Governments are increasingly giving up tasks that are genuine obligations in democratic states, namely the independent shaping of societal relations based on the democratic formulation of political demands and objectives, and decision-making. Instead, they are leaving it to international finance organizations and major multinational corporations to set the course for core issues in societal development.

The diminishing ability of governments to assert themselves is even more precarious wherever individual states are no longer singly responsible and problems must be solved in multi-lateral agreements or even by the international community of states. In this situation, national interests can be so conflicting, and the balance of power between industrial nations and the poorest countries so unequal, that effective and binding decisions—such as those needed to mitigate climate change—can hardly be agreed on and accepted.

All the same, globalization is neither a stroke of fate nor a natural law—it is expressly sought by the governments of the most important industrial nations. As one can guess, this is not entirely for unselfish reasons. Indeed it is quite in the national interests of these countries' powerful manufacturing and agricultural industries, as well as their trans-national service providers. So it is only understandable that the losers of globalization, or more precisely, ever larger segments of civil society affected by a globalization process that has rushed in completely un-invited and without their consent, feel threatened and resist the dictates of economics. These groups insist that at least the most important public goods be protected from commercialization. This applies especially to access to water because, in contrast to other goods, there is no alternative to water that consumers can switch to. There isn't any real competition between water suppliers either; whoever controls the mains, be they owner or licensee, has the monopoly.

Nevertheless, the call for a human right to water has its hidden dangers. For example, no one can seriously have a right to a full swimming pool. Nor can the irrigation of fruits for export, of tobacco, flowers, coffee, or vineyards, of fodder cereals for cattle, and of a hundred other luxury products be declared a human right. The state is obliged to guarantee a basic supply of drinking water to its population as cheaply as possible, but does this principle apply to the industrial sector as well? If so, to which products?

There are obviously different kinds of water, some of which are used for production purposes, while others meet basic, existential needs. Depending on the angle of approach, two different points of view are argued over. Access to water is broadly understood as a human right wherever debate centers primarily on existential needs, for instance at conferences on human rights, global development goals and sustain-ability, combating hunger and improving health, preventing the destruc-tion of the ecological balance and conserving biodiversity. But wherever water is seen as a commodity or commercial consumer item at con-ferences on economics and trade held by the World Trade Organization (WTO), the World Bank, the International Monetary Fund, or during WTO negotiations over the General Agreement on Trade in Services (GATS), human rights are considered little more than a disruptive factor distorting the free play of market forces. While negotiations over human rights establish standards, conferences on trade and economics bring into being measures which have real effect. Water issues will remain unresolved unless these two aspects are linked to each other.

Advocates of a human right to water rightly call for binding regulations such as an international convention on water. The proponents of such a convention aim at the safeguarding of water as a public good, making sure the human right of access has priority over international commer-cial law. Only state institutions can guarantee the human right to water. Human rights cannot be privatized. They are not commercial goods whose value is determined by the market. If such an international con-vention on water were ratified and put into force by the most prominent industrial nations, this would signify an important victory of a grand idea over reality.

Debate over climate change and other urgent issues facing humanity shows that the vessel of the community of states has a long braking distance, and that there are evidently numerous issues at present that many governments in the world regard as more important than human rights. But can the one-sixth of humanity who doesn't have access to safe water and sanitation facilities really wait until a water convention comes into effect perhaps twenty or thirty years from now?

Nevertheless, there is a good chance that considerable progress can be made in coming years. For one thing, the patent recipe favored by powerful international trade and finance institutions evidently hasn't yielded the success expected. Many privatization efforts have not met the expectations of the municipalities and states involved, nor of water companies and investors. International water businesses have greatly reduced their commitments, especially in developing countries. This may enable the development of new, indeed more modest solutions better adapted to local conditions.

For another thing, all 191 member states of the United Nations agreed to the millennium goals, pledging to make access to water one of their most pressing public responsibilities. A number of international agreements have already been made that directly or implicitly affect water management—agreements on protecting human rights, improving health, combating poverty and hunger, mitigating climate change, and encouraging sustainability.

One of the most important, internationally binding agreements, the 1966 International Covenant on Economic, Social and Cultural Rights, was ratified by 149 states, notably by the large industrial nations. The UN Committee on Economic, Social and Cultural Rights has written a General Comment to start making the unclear wording of this covenant more precise and close the loopholes that allow for loose interpretation. In coming years, it will establish standards regarding access to water that are increasingly binding and allow citizens to obtain rights through legal action.

This will set up a hierarchy for water usage which will differentiate between the imperative human right of drinking and maintaining hygiene, and, on the other hand, agricultural and industrial consumption, and other not strictly necessary purposes such as watering lawns, filling private swimming pools, or running fountains to decorate buildings and open spaces.

Objective standards are needed to determine the mandatory/obligatory measures states have to take to guarantee, as a public service, the water needed for existence. These standards can be summed up in four central points.

– Availability—are needed water resources available or can they be made available in a national effort or with aid from the international community?

– Accessibility–do these measures guarantee that the people affected will genuinely have access to water regardless of their financial means, race, or sex, and that no one, including children, people who are ill, and the elderly, will be discriminated against?

– Adaptability–are these measures relevant to a specific context and local conditions, taking economic, social, and cultural alternatives and special features into account?

– Acceptability–do measures take into consideration prevailing cultural and religious practices (and taboos) in each society?

This very pragmatic approach doesn't lose sight of the long-term goal, but does make sure the last step is not taken first. In practice, the intention of generally declaring access to water a human right can't be seen through without dealing with an enormous number of political, financial, economic, and legal complications. Added to that, the fight for protecting human rights is not at the top of the political agenda for most governments.

Nevertheless, this pragmatic approach puts words to demands that neither the political arena nor economic forces can ignore without being corrupted in face of the needs of people who are suffering. It lays down conditions under which water can be used as a commercial good and defines the minimum standards to be observed in supplying water to people. At the same time, it obliges governments to take definite action in accordance with internationally binding standards. It supplies governments and local authorities with needed methods and tools without prescribing specifically how these should be applied to meet goals.

This kind of strategy can also take into account local and regional differences, cultural traditions, and differences in prosperity between nations. Nevertheless, it lays down legally binding principles for resolving water conflicts between states. Finally, it also links human rights policies to economic and trade policies. It doesn't merely create a new set of papers and declarations–it establishes new facts and realities that guarantee every human on Earth the few liters of water he or she needs to live a dignified life.

But beyond the issue of the human right to water, there are a number of extremely dramatic ecological problems forcing governments to act. In many places, damage is already so enormous that it can't be played down by expressing doubt over scientific findings and publishing coun-terarguments. Unpopular and radical measures can't be put off any longer.

Many streams and rivers, especially in the Third World, with dozens of millions of people living along their banks, are so polluted that their waters are entirely unsuitable for any purpose. Four-fifths of China's rivers, among them the Yangtze and the Yellow River, and India's largest rivers, including the Brahmaputra, the Ganges, the Yamuna, the Goda-vari, and the Narmada have turned into brown, stinking, and foamy sewers in their lower reaches. Four out of ten rivers in the United States are so polluted that health authorities have issued warnings against swimming in them, not to mention drinking their waters.

Fossil groundwater reserves, a one-off emergency supply, have proven to be far less productive than expected in many regions with little rainfall. Many small aquifers, as well as some of the largest in the world, such as the Ogallala Aquifer in the United States, are slowly running low after a few decades of intense exploitation, a development that bears incalculable consequences for agricultural holdings and cattle farms. Groundwater levels are dropping meter after meter each year in many places in India and the Middle East, and in several countries in North Africa and Central America. Thousands of shallow wells, the basis of survival for many rural villages, are drying up as a consequence.

Our seas continue to be polluted with heavy metals, non-degradable chemicals, oil residues, pesticides, fertilizers, and untreated sewage. While the world's great oceans, Atlantic and Pacific, still aren't acutely threatened because of the immense volume of water they hold, the number of dead marine zones is going up steadily and now includes areas of the North Sea and Baltic, the Mediterranean, the Gulf of Mexico, the Persian Gulf, and the South China Sea. Many coastal waters, fishing grounds for millions of people, are seriously threatened.

Thanks to irrigation and the use of chemical fertilizers and modern agricultural technology, the living situation of several hundred million people in the Third World has greatly improved. But we are now seeing the dark side of the green revolution. Salinization and overfertilization have markedly lowered yields from much newly-won agricultural land after a few decades of intensive use. Each year more than one million hectares of cultivated land are irretrievably destroyed by salinization. Nearly one-third of all irrigated farmland is already damaged to some extent by salt in the topsoil. This trend will continue unabated unless radical measures are taken.

These findings are so overwhelming and alarming that governments, and more so, international organizations dealing directly or indirectly with these questions cannot overlook them. The failure of strategies pursued until now will force them, whether they care to or not, to intervene more decisively than ever before in the history of the world, despite the opposition of powerful business groups. The price of using water—for whatever purpose—must include the ensuing ecological and societal costs of repair. Water is a public good. Wherever businesses consume water, they must be obliged to use it in closed cycles or return it to the water cycle in the same condition they took it. Wherever this is demonstrably impossible, stringent environmental regulations must make sure that damage is kept to the lowest possible minimum. International agreements must protect rivers and seas, biodiversity and nature, and as such must be effective tools for enforcing policies on the management of chemicals and other pollutants, and on mitigating disruptive influences. If technical alternatives and environmental regulations are not stringent enough to effectively protect water, governments will be forced in the end to limit or ration its consumption.

In short, the sheer facts of reality will force upon us what gods and human-rights declarations apparently haven't managed to do, namely the insight that water is much more than a commodity to be used at will. Water as an irreplaceable element of life belongs to no one, but all of us together are responsible for it. It's all the same to water whether it's dirty, polluted, dammed, diverted, sold, or wasted, and it doesn't care who pretends to own it. But this can't be all the same to humanity, to us who need it.

洛 阳
LUOYANG

p. 1 Central Park, New York City, 1992. Thomas Hoepker/Magnum Photos
p. 2 Doug Hoke/Keystone
p. 4 Palm Beach, Florida. Eve Arnold/Magnum Photos
p. 6 Bangkok, 2005. Precha Keatchaithet/AP Photo
p. 8 Wolfgang Müller/Ostkreuz
p. 10 Martin Parr/Magnum Photos

p. 98 France. Harry Gruyaert/Magnum Photos
p. 100 Switzerland. Alessandro della Valle/Keystone
p. 102 Blueback salmon (Oncorhynchus nerka), Alaska. Stuart Westmorland/Keystone
p. 104 Yellowstone National Park, Wyoming. Fritz Pölking/WWF
p. 106 Brazil. Bruno Barbey/Magnum Photos
p. 108 Ian Gowland/Keystone/Science Photo Library
p. 110 Canyonlands National Park, Utah. Stuart Franklin/Magnum Photos

p. 518 Syria. Sylvain Grandadam/Keystone
p. 520 Seoul, South Korea, 2000. Yun Jai-hyoung/AP Photo
p. 522 Doonda, Barmer, India, 2000. Saurabh Das/AP Photo
p. 524 Windsor, Great Britain. Martin Parr/Magnum Photos
p. 526 Nanjing, China, 2006. Stringer/Reuters
p. 529 Doylestown, Pennsylvania, 2004. Rick Kintzel/Keystone

Index of countries, rivers, and lakes
Images in italics

Afghanistan *156*, 472
Alaska *298*, *299*
Algeria *124*, 135, 171
Amu Darya 190
Aral Sea 190–1, *192*, *194–5*, *196*
Argentina *220*, 248, 401–2, 430, 436, 442, 446, 463, 498
Armenia 336
Arno 472
Atlantic Ocean 516
Australia 371, 384–6
Austria *125*, *276*, 302, 394

Baghirati *360*
Bahrain 206
Baltic Sea 516
Bangladesh 227, 229, 247, 248, 252, 376, 399–401, 404, *478*
Banias 480–1
Barbados 170
Belgium 249
Benin 383
Bolivia 270, 378–9, 433–6, *438*, 442, 446, 460, 463
Brahmaputra 135, 150, 247, 515
Brazil *216*, 227, 235, 248, *309*, 314, 320, 337, 339, 356, 395, 402, 458, 500–1
Burkina Faso 383
Burma → Myanmar
Burundi 492

Calumet River 248
Cambodia *224*, *235*, *330*, *331*, 340, 341, 365–8, 493
Canada 289, 339, 372, 436, 446
Carmel River 374
Cauvery River 171
Chechnya 472
Chicago River 248
Chile *202*, 248, 372, 401–2
China 149, 153, *159*, *160*, 170–2, 191, 248, *280*, 284, *287*, 288–9, 294, *296*, *300*, *309*, 320, *322*, *324*, *326*, 337, 339, 341, 355–6, 365–8, 381 443, 451, *468*, *470*, 493–4, *526*
Colorado River 153, *161*, 189, 303, 489, 492
Columbia 436
Congo 227
Cuba *265*, *344*

Dan 480
Danube 302–3, 466, 494
Dead Sea 481
Denmark 170

Ebro 151
Egypt *124*, 135, *144*, *146*, 171, 484, 492–3
Elwha River 374–5
Equatorial Guinea *233*
Ethiopia 492–3
Euphrates 129, 472, 473, 476, *507*

Finland 474
France *138*, *139*, 228, 248, *298*, *299*, 302, 436, 441–3, 446, 451, 495

Ganges 135, 150, 152, 247, *360*, 515
Germany *242*, *244*, 247, 248, 268, *286*, *292*, 302, 304, *319*, 368–9, 394, *395*, 399, 404, 436, 443, 446, 451, *464*, 494–5, 500
Ghana 248, 382–3
Godavari 515
Great Britain *240*, 248, *264*, *283*, *310*, *318*, *319*, 386–8, *397*, 432, 443, 446, *448*, *524*
Guatemala 364–5
Guinea 383

Haiti *126*, *225*, *232*
Han 150
Hasbani 480, 481
Honduras 463
Hungary 303

Ili 191, *197*
Illinois River 248
India 135, 149, 150, 153, *159*, *160*, 170–2, 205, *214*, 227, *231*, 246–8, *262*, *266*, 269, 270, 288, 320, 337, 356, *360*, 366, 376–7, *389*, 392–3, 395, 444, 452, 454, 515, *522*
Indonesia *234*, 390–1, *428*, 436, 446, 450
Indus 128, 135
Iran 472
Iraq *186*, *232*, *456*, 472–7, *496*
Ireland 393–4, 499
Israel 135, 171, 172, 480–3, *506*
Italy *116*, *139*, 249, 451, 472

Jamaica *208*
Japan 150, 228, 290, 458
Jordan (river) 480–3
Jordan *200*, 480–3

Kazakhstan 190
Kenya 229, *230*, *246*, 268, 314
Kuwait 206

Lake Balkhash 191, *197*
Lake Havasu 303
Lake Michigan 248
Lake Pontchartrain 404, 407
Lake Tiberias 481
Lanzarote *202*
Laos *334*, 340–1, 365, 367, 493, 494
Lebanon *157*, 480–3
Libya 135, *164*, *166*, 171, 172, 206
Luxembourg 302, 495

Malaysia *225*, 356
Mali *118*, 229
Mauritius *264*
Mediterranean Sea 481, 516
Mekong 135, *330*, *331*, *332*, 340–3, 365–8, 493–4, *507*
Mexico 172, 227, 303, 337, 398, *484*, 489, 492, 502
Mississippi 248, 403
Morocco 135, *210*
Mozambique 436, 446
Myanmar *282*, 340–2, 366–7, 493–4

Narmada 150, 152, 515
Nepal *127*, *352*, 355
Netherlands 170, 302, 313, 404, 447, 495
Nicaragua 463
Niger *127*, *136*, *156*
Nigeria *222*, 227, *254*, 320, *448*
Nile 129, 135, *146*, 152, 341, 492–3
North Sea 516
Norway 337

Oman 370–1

Pacific Ocean 516
Pakistan 171, 229, 246–8, 320, 350, 356, 490
Palestine 203, 480–3, 506
Paraguay 337, 339
Paraná 339
Peru *234*, *353*, 356, 379
Philippines *131*, 172, *224*, 432–3, 442, 446, 463
Poland 289, 446
Puerto Rico 446, 463

Qatar 206, 463

Red Sea 481
Rhine 302, 369, 397, 495
Rio Grande 152, 489
Río Santa Cruz 402
Romania 303
Russia 249, 337

Sacramento River 374
Saint Lawrence River 289
Salween 342–3, 366, *507*

Saudi Arabia *120*, 135, 206
Scotland 454
Senegal *251*
Serbia 472
Slovakia 303
South Africa 314, 315, *377*, 447, 463, 501–2
South China Sea 516
South Korea 337, *520*
Songhua River 294
Spain 114, *138*, 151, 172, *202*, *299*, *369*, 371
Spree 248
Sri Lanka 395
Sudan *168*, 492
Sweden 314, 394
Switzerland 228, *272*, 302, *316*, *346*, 355, *358*, 397, 434, 451, 495
Syr Darya 190
Syria *125*, 473, 476, 480–3, *518*

Tajikistan *157*
Tapi 150
Thailand 172, 340–2, 365, 367, 493–4
Thames 248
Tibet 365, 366
Tigris 129, 472, 473, 476, *507*
Tisza 303
Tonle Sap *330*, 367
Tunisia 135, 336, 463
Turkey 337, 339, 473–7
Turkmenistan 190

Uganda 356, 492
United Arab Emirates *203*, 206
USA 114, 150, *154*, 162, 171–2, *174*, *176*, *180*, *181*, *184*, 206, 227, 247–9, 258, *260*, 268–9, *283*, 284, 289, *298*, *299*, 303, *308*, 337–9, 371, 373–6, 398, 402–7, 436, 441–3, 446, 451, 454, 482, 489, 492, 501, 516, *529*
Uzbekistan 190, *192*

Venezuela 339
Ventura River 374
Vietnam 248, *331*, 340–1, 355, 365, 367, 446, 472, 493–4

Xier 367

Yamuna 150, 246, 247, 515
Yangtze 128, 135, 149–50, 152, 289, 326, 515
Yarmouk 480–1
Yellow River 150, 152, *470*, 515
Yemen 172
Yucatan 395

Zarka 480
Zimbabwe 337

Further reading

Ball, Philip: H₂O. A Biography of Water.
London 1999

Barlow, Maude & Clarke, Tony: Blue Gold:
The Battle against Corporate Theft of the World's
Water. London 2003

Beach, Heather L. et al.: Transboundary Freshwater
Dispute Resolution. Theory, Practice, And Annotated
References. Tokyo, New York, Paris 2000

Clarke, Robin & King, Jannet: The Atlas of Water.
Mapping the World's Most Critical Resource.
Brighton 2004

Deckwirth, Christina: Sprudelnde Gewinne?
Transnationale Konzerne im Wassersektor und die
Rolle des GATS (Weed Arbeitspapier). Bonn 2004

FAO: World Agriculture: Towards 2015/2030.
Rom 2002

FAO: Review of World Water Resources by Country.
Water Report No 23. Rom 2003

McNeill, John R.: Something New under the Sun:
An Environmental History of the Twentieth-Century
World. New York 2001

Lanz, Klaus: The Greenpeace Book of Water.
London 1995

OECD: Environmental Outlook. Paris 2001

Shiva, Vandana: Water Wars: Privatization, Pollution,
and Profit. Cambridge 2002

Shiklomanov, I.H./Rodda, J.C.: World Water
Resources at the Beginning of the 21st Century.
Cambridge 2003

Stadler, Lisa & Hoering, Uwe: Das Wassermonopoly.
Von einem Allgemeingut und seiner Privatisierung.
Zürich 2003

UN/UNDP (Millennium Project): Health, Dignity,
And Development: What Will It Take? London 2005

UN/WWAP (United Nations/World Water Assess-
ment Programme): UN World Water Development
Report: Water for People, Water for Life. Paris,
New York, Oxford 2003

WHO/UNICEF: Global Water Supply and Sanitation
Assessment 2000 Report. Genf 2000
www.who.int/docstore/water_sanitation_health/
Globassessment/global2.1.htm

WHO: World Health Report 2002. Genf 2002

World Commission on Dams (Hg.):
Dams And Development: A New Framework
For Decision-Making. London 2000

Links

Centre for Environment and Development
for the Arab Regions und Europe (CEDARE):
www.isu2.cedare.org.eg

U.S. Environmental Protection Agency (EPA):
www.epa.gov

World Health Organization:
www.who.int/docstore/water_sanitation_health

Brot für die Welt: www.menschen-recht-wasser.de
(Information on international water companies,
multinational mineral water companies, etc.)

Public Citizen. Water for All Program
(Information on privatization, portrayals of
international water companies):
www.citizen.org/cmep/water

Fresh water, water resources, politics:
www.worldwater.org/index.html

Politics, conferences, etc.:
www.gdrc.org/uem/water/decade_05-15;
www.unescvo.org/water;

International River Network (NGO's, campaigns,
information on rivers): www.irn.org

History of wastewater treatment: www.wasser-
wissen.de/uebersichten/abwassergeschichte.htm

Editors

Klaus Lanz

Klaus Lanz was born in 1956. He is a water scientist and author. He studied chemistry and obtained his doctorate in Giessen, Germany, before doing environmental and water research at the Gray Freshwater Biological Institute at the University of Minnesota and the Swiss Federal Institute of Aquatic Science and Technology in Kastanienbaum-Lucerne. Klaus Lanz managed Greenpeace Germany's water campaign from 1988 to 1992 and is author of the Greenpeace Book of Water (1995). In 1995, he founded the independent institute International Water Affairs in Hamburg. He is investigating and commenting on water policy issues from a multi-disciplinary standpoint, with a focus on the interface of politics and science. He has worked with universities and research institutes, with environmental and development NGOs and with government agencies. Recently, he was a partner in the EU research project Watertime (2002–2005) exploring improvements of urban water management.

Lars Müller

Lars Müller was born in Oslo in 1955 and has been living in Switzerland since 1963. After doing his apprenticeship as a graphic designer, and years of apprenticeship and travel in the USA and Holland, he opened a studio in Baden in 1982. In 1983, Lars Müller began publishing books on typography, design, art, photography, and architecture.

Christian Rentsch

Christian Rentsch was born in Zurich in 1945. He studied German language and literature, as well as sociology and musicology in Zurich and Berlin. He has worked for more than thirty years as a freelance journalist, and as editor and department head on the cultural and media editorial board of the Zürcher Tages-Anzeiger newspaper. He lives in Erlenbach (near Zurich), where he works as a freelance journalist.

René Schwarzenbach

René Schwarzenbach, born in 1945, holds the position of full professor for environmental chemistry at the Department of Environmental Sciences at the ETH Zurich. He is currently head of the Department of Environmental Sciences and the school domain of Earth, the Environment and Natural Resources at the ETH Zurich. He is vice-president of Department IV of the Swiss National Science Foundation and a member of the advisory board of the journal Environmental Science and Technology. He obtained his doctorate at the Department of Chemistry at the ETH Zurich. Subsequently, he spent two years as a post-doctoral student at the Woods Hole Oceanographical Institute in the United States. In 1977, he accepted a position at the EAWAG, where he headed the department for interdisciplinary limnological research for several years and also served on the board of directors until April 2005. In 1992, René Schwarzenbach (along with four of his colleagues from Germany, France and Switzerland) was awarded the Körber Prize. In 2006, he became the first person outside the USA to receive the Award for Creative Advances in Environmental Science and Technology from the American Chemical Society. The textbook Environmental Organic Chemistry, 1994, which he wrote with two of his colleagues, is now considered a standard work on organic environmental chemistry.